Logic and Its Philosophy

Polish Contemporary Philosophy and Philosophical Humanities

Edited by Jan Hartman

Volume 11

UNIVERSITY of INFORMATION
TECHNOLOGY and MANAGEMENT
in Rzeszow, POLAND

PETER LANG

Jan Woleński

Logic and Its Philosophy

UNIVERSITY of INFORMATION
TECHNOLOGY and MANAGEMENT
in Rzeszow, POLAND

PETER LANG

Bibliographic Information published by the Deutsche Nationalbibliothek
The Deutsche Nationalbibliothek lists this publication in the Deutsche Nationalbibliografie; detailed bibliographic data is available in the internet at http://dnb.d-nb.de.

Library of Congress Cataloging-in-Publication Data
A CIP catalog record for this book has been applied for at the Library of Congress

This publication has been financed with the means of the Ministry of Science and Higher Education for maintaining the research potential of the Faculty of Administration and Social Sciences at the University of Information Technology and Management in Rzeszów, Poland

PETER LANG

UNIVERSITY of INFORMATION TECHNOLOGY and MANAGEMENT in Rzeszow, POLAND

Peter Lang GmbH Internationaler Verlag der Wissenschaften
Eschborner Landstrasse 42-50
60489 Frankfurt nad Menem
e-mail: frankfurt@peterlang.com
Tel.: +49 (0)69 78 07 05 0
www.peterlang.com

ISSN 2191-1878
ISBN 978-3-631-72745-4 (Print)
E-ISBN 978-3-631-72746-1 (E-PDF)
E-ISBN 978-3-631-72747-8 (EPUB)
E-ISBN 978-3-631-72748-5 (MOBI)
DOI 10.3726/b11395

University of Information Technology and Management in Rzeszów, Poland
Sucharskiego 2
35-225 Rzeszów
e-mail: wsiz@wsiz.rzeszow.pl
Tel.: +48 17 866 11 11
www.wsiz.rzeszow.pl

ISBN 978-83-64286-85-8 (Print)
E-ISBN 978-83-64286-86-5 (E-PDF)
E-ISBN 978-83-64286-87-2 (EPUB)
E-ISBN 978-83-64286-88-9 (MOBI)

This publication has been peer reviewed.

Table of Contents

Preface

This book collects 20 of my papers published in the years 2011–2016. All of them, with one exception, are devoted to logic and its philosophy. The essay XX is the only exception, although it also alludes to some logical questions. However, I included considerations on Keret's House for personal reasons. My roots are Jewish. I spent the years 1941–1944 in Warsaw, not in ghetto, but in relatively normal (if anything was normal at the time) circumstances. Although my family survived, but… (let me not finish). Hence, I am very sensitive to philosophical (and other) problems related to the Holocaust.

The rest of the book is a sequel to my *Essays on Logic and its Applications in Philiosophy*, published by Peter Lang in 2011. But this collection is more compact because all chapters belong to systematic philosophy (I did not include historical studies). However, I continue topics considered in the mentioned book of 2011 and use similar analytic tools taken from formal logic, especially the logical square and its generalization. Generally speaking, both collections can be viewed as contributions to so-called philosophical logic.

The papers included in this collection are reprinted here with changes introduced for avoiding repetitions. Yet, particular chapters overlap at some points; in particular, the diagrams re-appear in few places. Although I tried to unify symbolism to some extent, there are some differences caused by the fact that specific letters and signs play various roles. However, the context always explains what is going on at a given place. The numeration of formulas and statements is separate for each chapter.

The book is financially supported by the University of Information, Technology and Management in Rzeszów, Poland. I am grateful to its authorities, especially Dr. Wirgiliusz Gołąbek for this help. I would also like to thank Mr. Łukasz Gałecki for his initiative that this collection could be published by Peter Land. I am also grateful the text-editors for thei efforts toward preparing the final version of the book. All papers are reprinted with permission of particular publishers.

Jan Woleński

Source note

Semantic Loops, in Philosophy in Science. Methods and Applications, ed. by B. Brożek, W. P. Grygiel and J. Mączka, Copernicus Center Press, Kraków 2011, 256–271.

Logic as Calculus Versus Logic as Universal Medium, and Syntax Versus Semantics, *Logica Universalis* 6(3)(2012), 587–596.

Do We Need to Reform the Old T-Scheme, *Discusiones Filozóficas* (20)(13) (2012), 73–85.

Truth is Eternal if and only if it is Sempiternal, in: *Studies in the Philosophy of Herbert Hochberg*, ed. by E. Tegtmeier, Ontos, Frankfurt am Main 2012, 223-230.

Is Identity a Logical Constant and are there Accidental Identities?, *Studia Humana* 3-4(2012), studia humana.com/current-issue.html.com.

Naturalism and the Genesis of Logic, in *Papers on Logic and Rationality. Festschrift in Honor of Andrzej Grzegorczyk* (*Studies in Logic, Grammar and Rhetoric* 27(40)); together with K. Trzęsicki and S. Krajewski, University of Bialystok, Bialystok 2012, 223–240.

Some Analogies between Normative and Epistemic Discourse, in: *The Many Faces of Normativity*, ed, by J. Stelmach, B. Brożek and M. Hohol, Copernicus Center, Kraków 2013, 51–71.

Theology and Logic, in: *Logic in Theology*, ed. by B. Brożek, A. Olszewski and M. Hohol, Copernicus Center Press, Kraków 2013, 11–38.

Ens et Verum Concertuntur (Are Being and Truth Convertible)? A Contermporary Pesrpective, in: *Truth*, ed. by M. Dumitriu and G. Sandu, Editura Universitătii din Bucureşti, Bucureşsti 2013, 75–83.

An Abstract Approach to Bivalance, *Logic and Logical Philosophy* 23(1)(2014), 3–14.

Philosophical Reflections on Logic, Proof and Truth, in: *Trends in Logic XIII. Gentzen's and Jaskowski's Heritage. 80 Years of Natural Deduction and Sequent Calculi*, ed. by A. Indrzejcak, J. Kaczmarek and M. Zawidzki, Wydawnictwo Uniwersytetu Łódzkiego, Lódź 2014, 27–39.

Constructivism and Metamathematics, in *The Road to Universal Logic. Festschrift for 50th Birthday of Jean-Yves Béziau*, v. I, ed. by A. Koslow and A. Buchsbaum, Birkhäuser, Basel 2015, 513-520.

Rule-Following and Logic, in *Problems of Normativity, Rules and Rule-Following*, ed. by M. Araszkiewicz, P. Banas, T. Gizbert-Studnicki and Krzysztof Płeszka, Springer, Heidelberg 2014, 395–402.

Truth-Makers and Convention T, in *Philosophical Papers for Kevin Mulligan,* ed. by A. Reboul, in *Mind, Values, and Metaphysics. Philosophical Essays in Honor of Kevin Mulligan*, v. 1, Springer, Dordrecht 2014, 79–84.

Constructivism and Metamathematics, in *The Road to Universal Logic. Festschrift for 50th Birthday of Jean-Yves Béziau*, v. I, ed. by A. Koslow and A. Buchsbaum, Birkhäuser, Basel 2015, 513–520.

An Analysis of Logical Determinism, in *Themes from Ontology, Mind and Logic. Present and Past. Essays in Honour of Peter Simons*, ed. by S. Lapointe, Brill, Leiden 2015, 423–442.

Normativity of Logic, in *The Normative Mind*, ed. by J. Stelmach, B. Brożek and Ł. Kwiatek, Copernicus Center 2016, 169–195.

Formal and Informal Aspects of the Semantic Theory of Truth, in: *Uncovering Facts and Values. Studies in Contemporary Epistemology and Political Philosophy*, ed. by A. Kuźniar and J. Odrowąż-Sypniewska, Brill/Rodopi, Leiden 2016, 56–66.

The Paradox of Analyticity and Related Issues, in *Modern Logic 1850–1950. East and West*, ed. by F. Abeles and M. E. Fuller, Birkhäuser, Basel 2016, 135–138.

Some Liar-Like Paradoxes, *Logic-Philosophical Studies. Yearbook of St. Perterburg Logic Association* (14(2016), 70–75.

Nothingness, Philosophy and Keret's House, in: *Przestrzenie pustki/Void Spaces*, Fundacja Polskiej Sztuki Nowoczesnej/Foundation of Polish Modern Art., Warszawa 2013, 23–39.

I
Semantic Loops

Michael Heller introduced (see Heller 1999, pp. 90, 99–100) the idea of semantic loops. They are essentially involved in interactions between languages and their metalanguages (I will use the letter **L** as referring to a language and the symbol **ML** as denoting a metalanguage of **L**). As it is well-known, such interactions lead to semantic antinomies related to the self-referential use of expressions. The famous Liar antinomy (**LA** for brevity) is perhaps the most paradigmatic example. Consider the sentence:

(λ) the sentence denoted by (λ) is false.

A simple inspection shows that (λ) is true if and only if it is false (it will be demonstrated below). The situation illustrated by (λ) can be metaphorically characterized as a closed semantic loop, because we pass from truth to falsehood and back without any possibility of leaving the loop (or a circle, if you prefer this, more popular, figure of speech; see also a historico-terminological digression below). On the other hand, the language/metalanguage distinction cannot be liquidated, because we need to speak about languages and their various properties.

Although **ML** can be effectively reduced to **L** in the case of syntax (for example, via the method of Gödel numbering), this is impossible in the case of semantics; Tarski showed that doing semantics of **L** in **ML** requires that the latter is essentially richer than the former. Thus, closed semantic loops operate somehow between **L** and **ML**. As Heller indirectly (as he speaks about loops in physics) suggests, a solution of a semantic paradox assumes that loops related to such antinomies are not quite closed (for stylistic reasons, I use the labels 'antinomy' and 'paradox' as synonyms). In order to have a convenient *façon de parler*, I will contrast closed loops resulting in paradoxes and open loops, which save us from falling into antinomies as contradictions. The Gödelian sentence

(G) (G) is not provable,

provides an example of an open loop, because it is not paradoxical, but generates undecidable sentences or shows that arithmetic is incomplete (of course, after suitable formulation in terms of number theory and assuming its consistency); in fact, (G) is purely syntactic and does not involves semantic notions. My main task is to outline a formal approach to the problem of semantic loops.

Yet, I will also make some general and special philosophical suggestions at the end of this paper.

Let me start with a diagnosis of **LA** paradox outlined by Leśniewski and Tarski (see Tarski 1933, Tarski 1935, and especially Tarski 1944, pp. 671–674). They argued (Leśniewski never published his results) that **LA** is generated by the following facts (my formulation differs literally from that of Tarski):

(A) Self-referential application of 'true' and 'false' is admissible.
(B) The equivalence of the type (T) '*A* is true if and only if *A*' (the T-scheme; I use its naive form in this place) is accepted as the most basic formula explaining the concept of truth.
(C) Classical (bivalent) logic is employed in arguments.

This diagnosis opens three ways out. Firstly, we can exclude self-reference from semantics. Secondly, we drop the T-scheme. Thirdly, we decide to modify logic. In each case, a solution of **LA** costs something in the sense that we must sacrifice self-reference or the T-scheme or classical logic. Nothing is free of charge here and eliminating semantic antinomies requires a sacrifice of something for something. For Tarski (and the same concerns Leśniewski), playing with (A) was the most natural way (or cost less than rejecting the T-scheme) or changing logic. I do not know of any solutions consisting in rejecting the T-scheme. However, this remark must be properly understood. If we exclude self-reference, some T-equivalences resulting from the T-scheme are not admissible, for instance 'λ is true if and only if λ', because it leads to **LA** (see Thomason 1986). Similarly, we block analogous statements having the form of equivalences and related to other semantic paradoxes, for instance, Grelling's antinomy of heterological adjectives (see below).

It is important to see the role of the T-scheme in producing **LA**. Not every case of self-reference is harmful (in the sense of leading to an antinomy). In fact, we have 'innocent' self-referential sentences (for a general concise treatment of cases of this phenomenon, not only linguistic, see Smith 1986; Bartlett, Suber 1987 and Bartlett 1992 are useful collections) in which self-reference occurs. Consider, the phrase

(*) This sentence consists of 8 words.

Clearly, (*) is self-referential and false. However, it does not produce any antinomy. (T) serves as a device of abbreviation. Consider **LA** once again. The main step consists in observing that (a) λ = 'λ is false'. Hence, having (b) 'λ is true if and only if λ', and using (a), we arrive at: 'λ is true if and only if λ is false'. Similarly, define an autological adjective as such which possesses the property expressed by it; an adjective is heterological in the opposite case. For instance, the word 'short' is short and thereby autological, but the expression 'long' is just heterological. Now we ask: is 'hetelogical' heterological or autological? Denote 'heterological' by the symbol Ht and 'autological' by At. Assume that Ht is Ht, that is Ht(Ht) holds. If so, Ht is not-Ht, but At. In other words, if Ht(Ht), then At(Ht). Assume, then At(Ht). If so, 'heterological' possesses the property of being heterological, but this means that Ht(Ht) holds. Thus, Ht(Ht) if and only if At(Ht). Generally speaking, we have the equivalence (**) 'At(*E*) if and only if *E*(*E*)', which plays a similar role as the T-equivalence. Roughly speaking, semantic equivalences, like instances of (T) or (**), omit all terms expressing semantic properties, like truth or reference. Using the introduced terminology, semantic equivalences close loops.

Changing logic is a popular strategy in fighting against semantic antinomies. There are several attempts of this kind (see papers in Martin 1978 and Martin 1984, Sheard 1994, Antonelli 2000, Bromand 2001, Bolander, Hendricks, Pedersen 2006; this is only a sample of a very rich literature). Without entering into details, one may mention the transition from classical to many-valued logic (paradoxical sentences have another logical value than truth or falsehood; the most popular version assumes three-valued logic), gappy semantics (paradoxical sentences have no logical value) or paraconsistent logic (paradoxical sentences are so-called dialetheias, statements which are simultaneously true and false; λ is an example here). Other solutions propose to employ partial truth-definitions (it is a response to the undefinability truth-theorem; see below) or so-called circular concepts (see Gupta 2000), that is, concepts which are regulated by special revision processes. All mentioned ways out open semantic loops. I leave these proposals without further analysis, because I would like to show that the standard Tarski's recipe against antinomies or semantic closed loops is the simplest and most elegant solution.

I shall make one additional remark here. As I have already mentioned, any solution of **LA** and similar paradoxes comes at the expense of something. It is difficult to measure the amount which must or should be paid. Advocates of changing logic, partial truth-definitions or using circular concepts argue that we can define truth for **L** in **L** itself. They also point out that Tarski's solution via the hierarchy of languages is artificial and at odds with the spirit of natural language. However, it is difficult to understand what is behind such claims. Why should we prefer changing logic over a stratification of our language into levels? Although no conclusive or ultimate answer is available, the ordinary parlance suggests, for instance, in making citations, that we clearly distinguish between the language about things and the language about words.

Let me make a historico-terminological digression. The problem of circularity goes back to Aristotle. He pointed out that circular arguments and circular (or *idem per idem*) definitions are incorrect and should be eliminated. These issues belonged to elementary school logic for a long time. A new and important impulse came from mathematical logic. When Russell discovered his famous paradox of sets which are not their own elements, logicians began to propose various ways out. The vicious circle principle (**VCP**) was proposed as one of devices. It says that if a set X is defined, its elements cannot be characterized by belonging to X. Otherwise speaking, impredicative constructions should be abandoned. Unfortunately, this rule is inconsistent with some accepted definitions in the arithmetic of real numbers. Russell tried to harmonize **VCP** with mathematical practice via the ramified theory of logical types. However, this solution became contested, as it is enormously complicated. On the other hand, the role of circularity in the sense of **VCP** suggests that not every circular construction is bad. Perhaps, we can say that open semantic loops tolerate good circularity, but closed semantic loops generate paradoxes and must be broken. As far as I know, the terms 'loop' and 'closed loop' were proposed by Haim Gaifman (see Gaifman 1992, pp. 230–231, 253, Gaifman 2000, pp. 88, 109; see also Koons 2000, p. 192). According to Gaifman, loops are sets of unevaluated pointers, that is, items which point to sentences and determinate their semantics. Thus, loops in Gaifman's sense are something different than loops as understood in this paper, although the former are closely related to self-referential statements which generate **LA** and other paradoxes. Loops in Heller's (and mine) meaning are comparable with Gaifman's idea of black holes, that is, semantically gappy sentences. It is perhaps remarkable that various physical metaphors (loops, holes) became fairly popular in considerations about semantic difficulties stemming from self-referentiality. Let me add that Heller himself (see Heller 1999, p. 99) proposes a logic of loops as a weapon against closed loops and antinomies. According to him, such a logic should be non-linear, where non-linearity is understood analogically as in physics. As Heller points out to us, it is only a project. The name 'non-linear logic' can be slightly misleading, because the label 'linear logic' refers, in the jargon of contemporary logicians, to the logical theory combining classical and intuitionist logic in proof theory. Let me repeat, once again, that my further considerations are firmly set in the framework of classical logic and its metatheory.

As it is implicit in my previous remarks, Tarski's semantic paradigm plays a crucial role in this paper. His truth definition (see Tarski 1933, Tarski 1935, Tarski 1936) is the most spectacular example of the paradigm in question. The concept has been used earlier, but rather informally, due to **LA** and other paradoxes, which plagued logic in the first decades of the 20th century. Tarski defined truth formally. In the contemporary setting (I will employ it, not Tarski's original construction, except his remarks about the entire hierarchy of languages), truth is doubly relativized, namely to a language and a model. Thus, the fully defined context is 'true in a language **L** and a model **M**'. Let us assume that *A* is a sentence of a language **L**. We say that *A* is true if and only if it is satisfied by all sequences of objects taken from the carrier of **M** (equivalently: it is satisfied by one sequence; it is satisfied by the empty sentence). This definition obeys the convention **T**, that is, the condition:

(CT) A materially correct truth-definition logically entails all instances of the form:
(T) *n*(*A*) is true if and only if *A**, where '*n*(*A*)' is a name of the sentence *A* and '*A**' refers to its translation into a metalanguage (in fact, it is enough to include **L** into **ML**.

Here we have a new formulation of the T–scheme. As an example of a concrete T–equivalence, we have the sentence '*n*(snow is white) is true if and only if snow is white' for the sentence 'snow is white'. We assume that the expression '*n*(snow is white)' is a name of the sentence 'snow is white', and that the last expression is a self-translation into English as a metalanguage. Instead of writing '*n*(snow is white)' we could use quotes and say that the expression 'snow is white' is a name of the sentence occurring within the quotation marks. Also, instead of using English as our language and metalanguage simultaneously, we could say, for example, '*n*(Schnee is weiss)' is true in German if and only if snow is white, where German is our object-language and English serves as the metalanguage in which the mentioned equivalence is formulated. Remarks about using '*n*(…)' and quoting show that it is not easy to distinguish between naming sentences and using them (quotes indicate both these elements). In general, quotes (single '...', double "..." or any other kind, for example, ‹...›) are only conventional devices and can be replaced any other notation (this remark applies for others papers included in this book). Typically, labels for formulas are made in metamathematics by ascribing numerals for them; an example will be given below.

Perhaps, the most important feature of Tarski's (or semantic) truth-definition is its explicit metalinguistic character. Tarski originally worked in a version of Russell's theory of types. This framework forces an infinite hierarchy of languages $\mathbf{H} = \mathbf{L}_0, \mathbf{L}_1, \mathbf{L}_2$, ... such that $\mathbf{L}$ $\mathbf{L}_{n+1}$ (n = 0, 1, 2., ...) is $\mathbf{ML}_n$, that is, a metalanguage for the language $\mathbf{L}_n$. Consequently, we need a separate truth-predicate for every level. In fact, this variety is considerably limited, because it is cumulative: every next level contains former ones. The contemporary setting related to the distinction between first-order and higher-order languages assumes that it is enough to give the truth-definition for first-order sentences. If **L** is a first-order language, the truth-definition for its sentences is formulated in a second-order language as its metalanguage **M**; of course, we can always extend the construction for higher-order languages. Tarski showed that **ML** must be essentially richer than **L** in the sense that the former has to contain expressions of higher semantic categories than the latter and, moreover, the metalanguage has the translations of all expressions of the object language. Let us assume that **L** and **ML** have the same order. Then, we can construct a sentence, for example, λ, which is belongs to **L** and **ML** simultaneously and) leads to LA. This suggests that the set of true sentences in **L** cannot be defined in **L** itself. The last statement is a rough formulation of the Tarski undefinability theorem, one of the most important metalogical results. As far as the matter concerns the entire **H**, it is impossible to give a correct truth-definition for it. The Tarski undefinability theorem belongs to the family of limitative theorems, called in this way because they tell us something about limitations of formal systems. The Gödel incompleteness theorems, the Löwenheim-Skolem theorem about the size of models of first-order theories and the Church theorem about the undecidability of first-order logic are other members of this family. The first Gödel incompleteness theorem has two versions: syntactic and semantic. The former states that if Peano arithmetic (**PA**) is consistent, it is incomplete, because there are arithmetical sentences A and $\neg A$, such that both are not provable in **PA**, although the latter says that if **PA** is consistent, there are true but not provable arithmetical sentences.

The semantic version of the first Gödel theorem immediately suggests that there is a very deep connection between incompleteness and unprovability of truth in formal arithmetic. In my further remarks, I closely follow (as I did in Woleński 2004) the account given in Tarski, Mostowski, Robinson 1953, pp. 44–48 (I will also reproduce some proofs in order to demonstrate how formal devices work). Let **TH** be a first-order consistent formal theory with an infinite sequence of constant (that is, without free variables) terms $t_1, t_2, t_3, \ldots$. We say that a set of natural numbers $\mathbf{K} \subseteq \mathbf{N}$ is definable in **TH** (relatively to $t_1, t_2, t_3, \ldots$) if and only if there is a formula $A \in \mathbf{TH}$ such that $A(t_n)$ is valid in **TH**, provided that $n \in \mathbf{K}$ and $\neg A(t_n)$ is valid in **TH**, provided that $n \notin \mathbf{K}$. This definition is very easily generalized for arbitrary functions and relations. In particular, a function F is definable in **TH** if and only if for any $n \in \mathbf{N}$, the sentence $\forall v(A(t_n, u) \Leftrightarrow v = t_{Fn})$ is valid in **TH** and for any $m, n \in \mathbf{N}$ such that $Fm \neq n$, the sentence $\neg A(t_m, t_n)$ is valid in **TH**. Let us assume that we have any fixed one-to-one correspondence between natural numbers and expressions of **TH**; this strategy provides numerals as names of formulas. Without any loss of generality, we can concentrate on formulas of sentential category, that is, sentences or sentential open formulas. Thus, A_n is a formula corresponding with the number n and $Nr(A)$ refers to the correlated number of A. Now, we define d (that is, the diagonal function) by the equality $dn = Nr(A_n(t_n))$. Assume that v is the only free variable in A_n. According to the definition of the diagonal function, DFn is the correlated number of the formula obtainable from the formula $A_n(v)$ by replacing the variable v by the term t_n.

Let **V** be set of all natural numbers n such that A_n is a sentence true in **TH**. The following theorem holds:

(UT) If **TH** is consistent, then DG and **V** are not both definable in **TH**.

Proof: Assume that DF and **V** are both definable in **TH**. This entails that there are formulas B and C such that for any natural number n, (i) the sentence $\forall v(B(t_n, v) \Leftrightarrow v = t_{DFn})$ is valid in **TH** (the definability of d); (ii) $n \in \mathbf{V}$ if and only if $C(t_n)$ is valid in **TH** (by the definability of **V**). Since the variable v has no free occurrence in C, we can assume that it does not occur in C at all. Let $m = Nr(\forall v(B(v, w) \Rightarrow \neg C(v))$. This entails $A_m = \forall v(\underline{B}(w, v) \Rightarrow \neg C(v))$ and, furthermore, (iii) $A_m(t_m) = \forall v(B(v, w) \Rightarrow \neg C(v))$. If $A_n(t_n)$ is true, it is also true (by (i)) for the case $n = m$. This means that the sentence $\neg C(t_{dm})$ is also true. At the same time, however, provided that the sentence $A_m(t_{dm})$ is not true, then $Nr(A_m(t_{DFm}) \notin \mathbf{V}$. By the definition of the diagonal function, we have (iv) $dm = Nr(A_m(t_{dm})$. This implies (by the contraposition of (ii)) that $\neg C(t_m)$ is true in **TH**. Consequently, (v) the sentence $\neg C(t_{dm})$ is true in **TH**. The last result and (i) give (vi) $\forall v(B(t_m, v) \Rightarrow \neg C(v))$ is true in **TH**. Finally, using (i) and (vi), we obtain (vii) $C(t_{dm})$ is true in **TH**. Thus, we arrived at a contradiction, because $C(t_{dm})$ and $\neg C(t_{dm})$ are true in **TH**. Since we assumed that **TH** is consistent, the proof is concluded.

This result is followed by the following comment (p. 47): [This] theorem and its proof represent a metamathematical reconstruction and generalization of arguments involved in various semantical antinomies and, in particular, in the antinomy of the liar. The idea of this reconstruction and the realization of its far reaching implications is due to Gödel [...]. The present version of this reconstruction is distinguished by its generality and simplicity.

In fact, the set **V** is intended to cover the truths of arithmetic. Hence, the expression '∈ **V**' can read 'is true'; this justifies calling **V** 'the truth-predicate'. However, in order to avoid a confusion, let the symbol Tr stands for this predicate. Furthermore, the expression '$n \in$ **V**', as applied to sentences and due to the correspondence between natural names and sentences, can be read 'the sentence named by the number *n* is true'. Thus, the method of arithmetization provides a convenient system of naming sentences. If we look at (i) as applied to **V**, we obtain that

(CT') $\mathrm{Tr}(A_n) \Leftrightarrow A_{DFn}$,
holds for every sentence of **L**. This is, however, a version of the T-scheme. If we have
(FPL) **TH** $\vdash \mathrm{Tr}(A_n) \Leftrightarrow A_{DFn}$, for any $A \in$ **L(TH)**,

it means that **TH** (assuming that it is arithmetized theory) satisfies the fixed-point lemma (or the diagonalization lemma; see below) with respect to the predicate Tr; the expression A_{DFn} is the fix-point of $\mathbf{Tr}(A_n)$. Consider now the sentence $T_m = \neg\mathrm{Tr}(T_m)$. By (CT'), it is the fix-point of $\mathrm{Tr}(T_m)$. Thus, T_m is true, but on the other hand it is false by the assumption. This reasoning shows that the Liar paradox is a special case of the general proof of (UT). To see it, observe that (i) implies (CT') and the latter entails that we have the appropriate fix-points, depending on whether the correlated number, let say *k*, of the formula $\neg\mathrm{Tr}(T_m)$ belongs to **V** or not; thus, we consider the case when $k = m$. Thus, the formula (CT') $\Rightarrow \neg\mathrm{Tr}(T_m)$ (with T_m defined as above) becomes an instantiation of the formula $\forall v(B(v, w) \Rightarrow \neg C(v))$.

I will also give two other accounts of (UT). The first follows Bell 1999, pp. 219-222. Assume that $A_m(v)$ is a formula with the code number (the value of the diagonal function) *m*. Thus, for any *n*, it is possible to compute the code number of the sentence $A_m(\boldsymbol{n})$, where $\boldsymbol{n}$ is a numeral corresponding to *n*; the sentence $A_m(\boldsymbol{n})$ obtains from the open formula $A_m(v)$ by substituting the numeral $\boldsymbol{n}$ for the free variable *v*. This fact means that the computation in question is arithmetically representable and we are able to effectively define an arithmetical term $\mathbf{s}(u, v)$ such that for any natural numbers *m*, *n*, *p*, the sentence $\boldsymbol{s}(\boldsymbol{m}, \boldsymbol{n}) = \boldsymbol{p}$ is true if and only if *p* is the code number of the sentence $A_m(\boldsymbol{n})$. Under these preliminaries, we have:

(UT') If **V** is a set of arithmetical sentences satisfying the following conditions:

(a) Each member of **V** is true;
(b) The property 'the code number of a member of the set **V**' is arithmetically definable and there is a true sentence *G* of arithmetic such that $G, \neg G \notin$ **V**.

Proof. By (b) there is a formula $V(v)$ of **L** such that (c) $V(\boldsymbol{n})$ is true if and only if n is the code number of a sentence in **V** if and only if $A_n \in \mathbf{V}$, for every n. Consider the formula $\neg T(v)$. Substitute the expression $\boldsymbol{s}(v, v)$ for all free occurrences of the variable v in $\neg T(v)$. The resulting formula is $\neg T(\mathbf{s}(v, v)$. Denote it by $B(v)$. Suppose that m is its code number. Thus, B becomes the sentence A_m. Let p be the natural number such that the sentence $\boldsymbol{p} = \boldsymbol{s}(m, m)$ is true; thus, p is the code number of the sentence $A_m(\boldsymbol{m})$. Denote it by G. Therefore, p is the code number of G, that is, G is A_p. Observe now that we have the sequence of equivalences:

(E) G is true iff $A_m(\boldsymbol{m})$ is true iff $B(\boldsymbol{m})$ is true iff $\neg T(\boldsymbol{s}(\boldsymbol{m}, \boldsymbol{m})$ is true iff $T(\boldsymbol{p})$ is false if $A_p \notin \mathbf{V}$ and if $G \notin \mathbf{V}$.

(E) forces the conclusion that G asserts of itself that it does not belong to **V**. We also conclude that G is true, because, in the opposite case, it would be true as a member of **V**. Thus, G is true and $G \notin \mathbf{V}$. It implies that $\neg G$ is false and it does not belong to **V**, since this set consists of true sentences only. This completes the proof of (UT'). If **V** is taken as the set of all arithmetical truths, we obtain (AUT), The property being a true arithmetical sentence is not arithmetically definable. A terminological comment is in order. Bell labels (UT') as the arithmetical truth-theorem and he calls (AUT) the Tarski undefinability theorem. This seems not quite proper, because (UT) was proved by Tarski and his collaborators for all arithmetizable theories. Thus, (UT') is just a counterpart of (UT) in Bell's language and (AUT') is the truth undefinability theorem for arithmetic.

The second account (see Smullyan 1992, p. 104, Murawski 1999, p. 204) is the simplest one. First, we observe that the fixed-point lemma holds for **PA** (the symbol **PA** refers to Peano arithmetic), that is, we have:

(FPLPA) **PA** $\vdash \mathrm{Tr}(A_n) \Leftrightarrow A_{dn}$, for any $A \in \mathbf{L}(\mathbf{PA})$,

Assume that there is a formula $A(v)$ such that for any sentence S of **L(PA)** **PA** $\vdash S \Leftrightarrow A(S_{dn})$. By (FPLPA) there is a sentence S' such that **PA** $\vdash S' \Leftrightarrow \neg A(S'_{dn})$. This immediately leads to a contradiction, because we conclude: **PA** $\vdash A(S'_{dn}) \Leftrightarrow \neg A(S'_{dn})$. Thus, there is no formula of **L(PA)** such that it is the truth-definition for **PA**. It is, of course, the theorem of the undefinability of truth for **PA**. It can be generalized for any theory, which contains **PA** as its subtheory. Suppose that the notion of proof is formalized and we have (*) if every member of a set **X** of sentences is true, then all sentences provable from **X** are true; (**) if the property of being a member of the set **X** is arithmetically definable, then the property of a sentence provable from **X** is also arithmetically definable. Let **X*** be the set of all sentences provable from **X**, provided that the latter consists of truths. Thus, by (*) and (**), **X*** consists of true sentences and the property of being its member is arithmetically definable. This allows us to apply (**UT'**) and conclude that there is a true sentence G such that neither it nor its negation belongs to **X***. In other words, we have the first Gödel incompleteness theorem:

(GT1) There are true and unprovable arithmetical sentences,

in the semantic setting (Bell calls it "Gödel's weak incompleteness theorem"). Its syntactic (or strong) formulation requires much more advanced methods. In particular, the assumption of the truth of **X** is replaced by the supposition that this set is ω-consistent (or even consistent after Rosser). The advantage of the strong version is that its proof provides an example of an unprovable sentence, but the weak version only states that such sentences exist. However, the sentence G asserts its own unprovability in both versions, although G used in the proof provides a of (UT') expresses its own falsehood. The limitations asserted by (UT) and (GT1) are not accidental. Although they can be eventually repaired by adding the ω-rule as a new inference rule or admitting that the considered theories have non-recursive axiomatizations, these moves demolish the finitary character of deductive means.

I shall now return to the fixed-point lemma (theorem). In mathematics, for instance, in topology, we frequently encounter the situation that $fx = x$, where f is a function. If it is the case, we say that x constitutes a fixed point of fx. The equality sign in the formula $fx = x$ admits a variety of interpretations, including its reading as ⇔. In fact, (FPL) and (FPLPA) employ just this interpretations. Roughly speaking (see Smullyan 1992, pp. 102–103), (FPL), as a general metamathematical theorem, says that, for any formula FA asserting that A satisfies the clause F, FA has a fixed point A, that is, the equivalence $FA \Leftrightarrow A$ holds, provided that the diagonal function is definable in a given theory. If d is definable in **TH**, which is a formal theory (for a survey of fixed-point constructions in logic, see Sommaruga-Rosolemos 1991), this means that **TH** $\vdash d(A) \Leftrightarrow A$. In particular, **TH** $\vdash d(\lambda) \Leftrightarrow \lambda$, which leads to LA. It is interesting that we do not need to invoke self-reference directly. The fixed-point lemma and diagnonalization in formalized theories act similarly to self-reference in natural language, at least as far as the matter concerns general effects. Instead of eliminating self-referential context, we must decide whether to produce all possible **T**–equivalences (define the function d as total) or to define the set **V** of true sentences. Combining both these tasks inevitably leads to antinomies. Otherwise speaking, the choice is either a partial truth-definition for a given **L** with the truth-predicate belonging to this language or the full truth-definition but formulated in **ML**. If (CT) is a condition for a correct truth-definition, it blocks maximalism consisting in defining **V** and

d together; compare the content of (UT). Clearly, the introduced formal machinery refines Tarski's analysis of natural languages.

Stephen Yablo (see Yablo 1985) argues that Tarski's original approach does not disprove all versions of **LA**. Before presenting Yablo's reasoning, I shall recall another version of **LA** produced by the pair of sentences A_1: A_2 is true, A_2: A_1 is false. Assume that A_1 is true. Thus, A_2 is true as well, but A_1 appears as false, contrary to the assumption. Assume that A_1 is false. This entails that A_2 is false and, then, that A_1 is false, once again contrary to the adopted assumption. The circle of liars (or the loop of liars) shows that self-reference can be indirect. In fact, none of $\{A_1, A_2\}$ is self-referential in itself, but it is only their combination that has this property. Yablo proposes to consider an infinite sequence S of sentences: $A_0, A_1, A_2, ..., A_i, ...$, where A_0 = 'A_k is false for any $k > 0$', A_1 = 'A_k is false for any $k > 1$', A_2 = 'A_k is false for any $k > 2$', ..., A_i = 'A_i is false for any k > i', ... Fix an arbitrary i. By assumptions, A_i is false for all $k > i$. When we then apply the T-scheme, we obtain that (α) A_i is true if and only if A_i is false for all $k > i$. Assume that A_i is true. This entails that A_{i+1} is false, A_{i+2} is false, etc. Now, if A_{i+1}, then A_{i+2} is true, contrary to the right part of (α). The last assertion entails that A_i is false for all $k > i$. Consequently, the left side of (α) is false as well, contrary to our assumption. The supposition that the formula 'A_i is true' is false immediately leads to contradiction, because it entails that at least one A_k is true.

Yablo labels the paradox constituted by S as the ω-Liar and claims that it has nothing in common with self-reference and language-hierachies, even indirectly. More precisely (see Yablo 2006), he argues that the typical strategy of anti-circularity recommended by the Tarskian-style approach does not apply to the ω-Liar. Regardless of Yablo's own recipes, his critical remarks against Tarski are not quite correct. Consider the sequence S once again. He has the stock of sentences $A_0, A_1, A_2, ..., A_i$ and the condition that for any $k > i$ ($k = 0, 1, 2, ..., i, ...$), A_i asserts that A_k is false. A consistent description of S requires that A_0 is true, because in the opposite case there is at least one A_k which is true. Thus, we can conclude that A_o indirectly asserts its own truth. Similarly, we argue that since no sentence A_i ($i > 0$) can be true, each of these sentences indirectly refers to its falsehood. A more important observation points out that since the ω-Liar heavily uses the T-scheme, it depends on the (FPL). Consequently, the ω-Liar shows that the set of true sentences is not definable in the language in which this antinomy is formulated. Thus, diagonalization and (FPL) appear as much more fundamental phenomena than self-reference.

What about self-reference to truth? The sentence

(τ) the sentence denoted by (τ) is true,

is called the Truth-Teller. It is self-referential and informs about its truth; (τ) expresses. Is (τ) paradoxical? Common wisdom says: No. The justification is as follows. (FPL) produces the formula (β) Tr (τ) ⇔ τ as provable in arithmetic. Moreover (see Löb 1955), if τ is provable, it is true. This fact does not lead to an antinomy and is consistent with (UT). The equivalence (β) expresses one of the so-called reflection principles; it reflects about itself, its own truth in this case. However (see Woleński 1995 for details), we can formulate the paradox of Truth-Teller by assuming the rejection logic or dual logic as our formal frame. Roughly speaking, classical logic as understood in (C) distinguished truth as a logical value. This means that the consequence operation associated with logic preserves truth and formalizes the process of assertion of sentences on the basis of other sentences, already asserted. However, nothing, at least from a purely formal point of view, prevents using logic which preserves falsehood and operates on rejected sentences. For example, if A or B is rejected, then $A \wedge B$ is rejected; if $A \vee B$ is rejected, A is rejected and B is rejected. Both these rules are formulated in the metalanguage of the assertion logic. Yet, we can build the rejection logic in **L** and **ML** without referring to the language of assertion. In this new formal framework, which has the duals of all standard metamathematical results (λ) is harmless, but (τ) leads to a paradox. This means that we have no general criteria to decide *a priori*, which cases of self-reference are innocent and which are not. An accessible empirical evidence suggests that dangerous self-referential cases are semantic in their essence (they involve categories like truth, reference, etc.) and occur together with reduction schemes as the T-scheme allowing a compression of self-referential sentences. Once again, the functions *d* and (FPL) play a crucial role. In fact, they liberate us from appealing to self-reference, at least in well-(formally)behaved systems. Self-reference becomes essential, when we pass to natural languages.

My previous considerations suggest that the assertion "semantic loops are created by self-reference" should be taken *cum grano salis* and only as the first approximation. A closer analysis via metamathematical devices shows that diagonalizations and (FPL) produce such loops (also syntactic) in well-defined formal theories. Self-reference appears when we work inside natural languages in which such concepts as truth, provability or definability function as not quite determined categories. At the same time, however, there is a fairly close correlation between both sources of loops, that is, precise mathematical construction and self-referential contexts, because both produce inconsistencies, unless some restrictions are imposed on the performed constructions. Returning to the beginning of this paper, we should observe that some loops are open, but other closed.

By definition, open loops are consistent, whereas closed ones result in antinomies. Hence, we need a way to open closed loops in order to save our systems from inconsistencies. Tarski's idea was to introduce language hierarchies and eliminate self-referential sentences. The claim that concepts like satisfaction, truth or denotation should be defined in metalanguages became a by-product of this strategy. Some people consider this device as artificial and propose various alternative methods (see above; see also, for instance, Kripke 1976, Barwise, Etchementy 1987, Barwise, Moss 1996), including strategies employing fixed-points constructions, although combined with reforms of logic or set theory. I would like to propose another look at opening closed loops, which, at least in my opinion, is a straightforward application of Tarski's way of thinking.

Let us examine (UT) once again. It says that the diagonalization function and the set of truth are not together definable in a consistent, rich formal system (containing arithmetic). We can say that diagonalization produces a semantic loop, which is closed by any truth-definition satisfying (CT). The way out is simple: resign from defining truth for **L** in **L** and pass to **ML**. In other words, (UT) opens the closed loop in question. Similarly, the incompleteness theorem opens the closed loop produced by the claim that arithmetic is complete (or finitely axiomatizable). Both limitative results save consistency. The difference between the "Tarskian loop" and the "Gödelian" loop consists in the fact that although the latter can be opened in **L**, the former cannot. Simply speaking, this fact reflects a very deep difference between syntax and semantics. The very amazing result, which formally exhibits this situation (see Murawski 1999, pp. 284–285), states that the concept of truth does not belong to the arithmetical hierarchy, but it belongs to the analytical one, although the notion of provability can be incorporated into the former. Due to the well- known features of natural languages, we cannot apply *DG* or (FPL) to them, even indirectly. However, we can say that (CT) determines our use of the predicate 'is true' in the ordinary speech and every correct truth-definition should imply all instances of the T-scheme. Since it is demonstrable that some well-formed ordinary sentences lead to antinomies, something must be done in order to avoid inconsistency. We claim that truth is undefinable and that the hierarchy of language should be introduced. The advantages of the proposed picture are as follows. Firstly, the strategy is very simple, simpler than its alternatives. Secondly, self-reference becomes de-mystified (see also Smorynski 1985, Chapter 0). Thirdly, self-referential assertions do not need to be treated as ill-formed or ungrammatical.

It remains to touch on some general philosophical points suggested by the foregoing arguments. I deliberately abstain from saying that formal logical constructions entail (in the strict logical sense) definite philosophical consequences, because, according to my view, the relation between sciences (including logic and mathematics) is very complex and requires much reflection. However, let me speculate on this issue. Assume that we have a language **L** suitable to express the entire knowledge in it. This means that **ML** is a part of **L**. Due to (UT) the set of true sentences of **L** has no definition in this language. Since no other language is available, the extension of 'is true in **L**' cannot be capture by any **L**-formula. It appears natural to say that the concept of being corresponds to the set of all truths of **L**. However, the undefinability of truth implies the undefinability of being. These assertions agree with the theory of transcendentals developed by medieval philosophers (see Woleński 1997). The next philosophical observation refers to the nature of deductive devices used in metamathematics. Although our actual deductive capabilities are finitary, arithmetic can be made complete by adding the ω-rule, which is infinite as far as the matter concerns the cardinality of the set of its premises. Of course, due to our deductive nature, this does not mean that we are able to use the ω-rule effectively, although we can speak about its effects. At the same time, however, (UT) cannot be overcome by supplementing our deductive machinery by infinitary devices. Otherwise speaking, the concept of truth (and being *a fortiori*) remains undefinable after adding infinite rules of inference. This remark suggests a modesty in speculations about epistemic potentialities of "infinite minds", very frequently offered by theologians. The same warning concerns theories of everything.

Heller (see Heller 1999) makes some more general suggestions concerning loops (not necessarily semantic) as occurring in physics. This direction seems interesting and deserves a continuation. I would like to offer a small contribution to this line of thinking. The semantic truth-definition (see above) defines 'is true in **L** and **M**.' These relativizations are frequently considered as artificial and at odds with our intuitions. However, think about **L** and **M** as semantic frames of reference, similar to those used in physics, for example, in the special theory of relativity. The claim that truth should be defined without any reference to languages and models is similar to the pretension that time flows absolutely and the same way for arbitrary frames. More importantly, the relativizations used in the semantic analysis of truth do not deprive this concept of the mark of objectivity, just as time is an objective category of physics. In fact, physical loops and semantic loops provided some theoretical constraints. They remain open, although the observers and observables interact in both cases. Metaphorically speaking, defining truth for **L** in **L** without additional postulates generates a "semantic speed" that exceeds "the speed of semantic light".

II
Logic as Calculus vs. Logic as Universal Medium, and Syntax vs. Semantics

The distinction between logic as calculus (**LoC**) and logic as language (**LoL**) was introduced by van Heijenoort (see Van Heijenoort 1967). He regarded this distinction as a very useful device for the history of logic, closely resembling Leibniz's famous pair of *calculus ratiocinator* and *lingua characteristica.* According to van Heijenoort, logic under Frege's understanding was rather a kind of *lingua characteristic*, contrary to the view of Boole or Schröder seeing logic as a calculus. Russell and Whitehead followed Frege's view, making some changes forced by the theory of logical types. A new tradition, more related to Schröder than to Frege, was initiated by Löwenheim in 1915 (see also Badesa 2004), who proved the first model-theoretic theorem (if a formula of first-order logic is valid in a denumerable domain, it is valid in every domain). Generally speaking, the opposition between **LoC** and **LoL** was replaced by the axiomatic approach to logic and its set-theoretic account. Skolem, Gödel and Herbrand made essential contributions to this new world of logic. In particular, van Heijenoort finds Herbrand's work as a possible third way. Van Heijenoort seems to consider **LoC** and **LoL** as ideal types. For instance, he stresses that Frege characterized logic not only as a *lingua characteristica* but also as a *calculus ratiocinator.* Although van Heijenoort does not give any general characteristics of both types of logic, he points out some of their important features, particularly of Frege's conception. For Frege, logic is universal and concerned the stable (not changeable) universe. Since logic functions as the universal language, there is no stance outside of it. Hence, no metalogical problems can arise, because it would result in transcending the language limits, which is, by definition, impossible.

Van Heijeenort's conceptual dichotomy of **LoC** and **LoL** was generalized by Hintikka (see Hintikka 1989, Hintikka 1997) and Kusch (see Kusch 1989), who introduced and elaborated the distinction between language as a calculus (**LaC**) and language as a universal medium (**LaM**). Kusch (Kusch 1967, pp. 6–7) proposes the following characterization:

LaC	**LaM**
1. Semantics is accessible.	1.' Semantics is inaccessible.
2. Different systems of semantic relations are conceivable.	2'. Different systems of semantic are inconceivable.
3. Model theory is accepted.	3'. Model theory is rejected.
4. Semantic Kantianism is rejected.	4'. Semantic Kantianism is adopted.
5. Metalanguage is not legitimate.	5.' Metalanguage is legitimate.
6. Truth as correspondence is intelligible.	6'. Truth as correspondence is not intelligible.
7. Formalism is linked with the thesis 1.	7'. Formalism is linked with the thesis 1'.

Clearly, since language cannot be reduced to logic, **LaC** generalizes **LoC** and **LaM** provides a generalization of **LoL**.

The pair <**LaC**, **LaM**> enlightens some aspects of the history of logic and philosophy in the 20th century (see Hintikka 1988, Woleński 1997a). Apart from Frege, Wittgenstein was perhaps the purest representative of **LaM**. The statements 5.6 (5. 6: *Die Grenzen meiner Sprache* beduten dir Grenzen meiner Welt – *The limits of my language* mean the limit of my world) and a part of 5.61 (Die Logik erfüllt die Welt; die Grenzen der Wlt sind auch ihre Welt – Logic fills the world: the limits of the world are also its limits) in Wittgenstein 1922 are probably the most extreme expressions of semantic Kantianism. Works by Husserl, Carnap and Tarski, in turn, can be interpreted as representative examples of the **LaC** conception. In general, Hintikka identifies **LaC** with the model-theoretic tradition in logic. Although I agree that both distinctions in question considerably help us to gain a better understanding of various crucial episodes of contemporary philosophy – both analytic and continental – I claim that they are less important than their proponents think. I shall restrict my remarks to the philosophy of logic, because any attempt to investigate the role of language in the entire philosophy (or even in its part) exceeds the scope of a single paper. The main device I apply is the distinction between syntax and semantics which I believe to be the most important tool for any correct account of the nature of logic. More specifically, I will argue that logic, including metalogic (metamathematics), is both a calculus and a language. Rephrasing Kant's famous dictum, syntax without semantics is empty, but semantics without syntax is blind. Or, using van Heijenoort's distinction between logic as a calculus and logic as a language, we can say that the former is empty without the latter and **LoL** is blind without **LoC**. Yet, this colorful framing of the problem does not entail semantic Kantianism.

It is intriguing that van Heijenoort mentions Hilbert only occasionally (Van Heijenoort, p. 15, footnote 6):

> The notion of formal system was again bought into the forefront by Hilbert, in the 1920s.

The word 'again' suggests that Hilbert followed someone else in bringing the concept of formal system into the forefront. In fact, van Heijenoort to some extent considers Hilbert as a follower of the axiomatic approach to logic initiated by Frege and Russell. Hintikka ascribes much more importance to Hilbert than van Heijenoort does (see Hintikka 1988, pp. 108–110) and says that the founding father of formalism in the foundational studies mastered the axiomatic approach to logic and mathematics. According to Hintikka, Hilbert's axiomatics was an essential contribution to the model-theoretic tradition in logic. However, although, pace Hintikka, Hilbert was very far from representing a "blind" formalist, i.e. a philosopher of mathematics neglecting the content of formal systems, his main merit for the foundations of mathematics consists in initiating the systematic metamathematics, that is, mathematical investigations of formal systems. Thus, the problems of consistency and completeness became mathematical problems due to Hilbert's foundational program (see Sieg 2000 for a survey of Hilbert's way toward finitary proof theory, which is the essence of metamathematics).[1] It is legitimate to consider the finitistic approach to formal systems from the perspective of the finitist proof theory as syntactics. Thus, for Hilbert a formal system is a purely syntactic object entirely defined by its vocabulary (a stock of symbols) and the recursive rules which prescribe how to construct complex strings of symbols from the initial atoms. Such a picture of formal systems perfectly fits the idea of **LoC** and **LaC.**

Even if we agree that Hilbert's main contribution belongs to the formalist tradition in the foundations of mathematics, we should remember that he used semantic concepts, like domain or satisfaction, in an informal way. Since I identify the model-theoretic tradition with the semantic tradition, the rise of formal semantics is a crucial fact for this perspective. There are currently two uses of the term 'formal semantics' (I follow Woleński 2004a). The first, becoming more and more popular among linguists, refers to a formal (mathematical) approach to natural languages (see Cann 1983, pp. 1–2) and their relation to the world.

1 Emil Post's proof of the completeness of propositional calculus in 1920 (see Post 1921) is a phenomenon deserving a separate historical explanation. All references in Post's paper indicate that he was exclusively inspired by *Principia Mathematica* by Russell and Whitehead; Hilbert is not mentioned by Post.

Formal semantics in this meaning accepts the following theses: (a) Chomsky's thesis that natural language is a formal system; (b) Montague's thesis that there is no principal difference between formal and natural languages. However, the real scope of the validity of (a) and (b) is still disputed. Hence, it is not clear to what extent speaking about the formal semantics of natural languages is justified. Since I am not going to discuss this question, let me now pass to the second meaning of formal semantics, according to which it consists in a mathematical approach to formal languages and their semantic features, such as truth, reference, etc. Formal semantics in this sense is the same as model theory. Roughly speaking, to give a semantic characteristics of a language **L** means to point out the class **C(M)** of models of **L**, i.e. the class of algebraic structures of **L** such that the sentences of **L** are true in every **M** belonging to **C(M)**. Thus, we have **L** as a formal language and **C(M)** as a formal set-theoretical object. Since both are mathematically characterized, everything is formal in formal semantics of formal languages.

However, the outlined picture is oversimplified. The following opposites applied to languages are relevant to our problem: (A) natural – artificial; (B) informal – formal; (C) unformalized – formalized; (D) interpreted – uninterpreted. At first glance it seems that members of the sequence <natural, informal, unformalized, interpreted> express the same property, which can also be attributed to 'ordinary', 'colloquial', etc. Consequently, the words 'artificial', 'formal', 'formalized' and 'uninterptreted' appear to indicate the same feature, contrasted with the property expressed by such terms as 'natural', 'informal', etc. A closer inspection shows that these extensional identifications are dubious. If we say that a language is artificial, we mean that it was created to perform some specific tasks. Although an artificial language contain special symbols, it does not need to be formalized. According to a basic intuition, a language is formal if it can be described independently of the content of its expressions and appeals only to their syntactic form. Finally, a formalized language is a result of a special process or operation, called formalization. Nothing precludes that a formalized language arose as a formalization of an informal one. For instance, formal mathematics arises a a result of formalization of more or less informal mathematics. Furthermore, nothing precludes a formalized language from being interpreted. This was strongly stressed by Tarski (see Tarski 1956, pp. 166–167; the work was published in Polish already in 1933):

> It remains perhaps to add that we are not interested here in 'formal' languages in sciences in one special sense of the word 'formal', namely sciences to the signs and expressions of which no material sense is attached. For such sciences the problem here discussed [the problem of truth] has no relevance, it is not even meaningful. We shall

> always ascribe quite concrete and, for us, intelligible meanings to the signs which occur in the language we shall consider. The expressions which we call sentences still remain sentences after the signs which occur in them have been translated into colloquial language. The sentences which are distinguished as axioms seem to us materially true, and in choosing rules of inference we are always guided by the principle that when such rules are applied to true sentences the sentences obtained by their use should also be true.

For Tarski, doing formal semantics for **L** requires that the language in question is interpreted, although formalized. Tarski himself offers a very general description of formalized languages as those in which the sense (meaning) of every expression is "unambiguously determined by its form" (Tarski 1956, pp. 165–166). On the one hand, this account certainly leaves important questions open. In particular, one can ask: what is the sense? On the other hand, he rather relied on carefully listed features of formalized languages; for instance, the fact that all constructions in such languages are performed according to structural properties and rules.

Perhaps, it is the most important outcome of the foregoing discussion is the conclusion that such features as formalized and interpreted can coexist. This raises the question of how the interpretation of a formalized language **L** (for simplicity, I assume that **L** is a first-order language without functions, symbols and the identity predicate) proceeds. The first step to describe **L** is to specify its alphabet. Without going into details, the alphabet of **L** contains logical constants (propositional connectives, quantifiers), individual variables and constants, and predicate letters. The set of formulas of **L** includes atomic items of the type $P(t_1, \ldots, t_n)$, where P is a predicate letter and $t_1, \ldots, t_n$ are terms (individual variables or individual constants) and compound formulas built from connectives and quantificational closures, for instance $\neg Px$ or $\exists xPx$. In order to speak about semantic relations, we have to define an interpretation of **L**, i.e. to specify an interpretation function **v**. The first step consists in choosing the universe **U** fixing the set of values for individual variables and the scope of quantifiers. The next step is to define the function **v** (the valuation function) which ascribes determined elements of **U** to individual constants and denotations (subsets of **U** or relations defined within **U**; I will call them attributes expressible in **L**) of predicate-letters. Models of **L** are those interpretations in which sentences of the language in question are true.

It is not true that a purely formal language (at least, as defined here) does not yield any information about its possible models. Every language (more precisely, its alphabet) has its signature generated by the arity of its individual constants and predicates. Generally speaking, the arity indicates the number of arguments. By definition, the arity of constants is equal to 0, the arity of monadic predicates is equal to 1 and the arity of dyadic predicates is equal to 2. Since the signature of the alphabet is unique, it also determines some structural properties of interpretation. If the signature of **L** is <0; 1, 2>, we know in advance that any interpretation of **L** must contain at least one distinguished object (since **v** is not a bijection, more than one name can refer to the same object) and two attributes, one monadic and one dyadic. However, connections between languages and their possible interpretative structures generated by the signatures do not yield interpretations. Although the function **v** interprets the alphabet, it suffices to interpret the whole **L**, provided that **L** is semantically compositional. The compositionality means here that the semantic value of a complex expression is the value of a function, denoted as **v**, of the semantic values of their explicit components. This is the case if composite expressions are formed by applying syntactic rules to already constructed items. One could maintain that the requirement of compositionality is too strong, as such a claim creates difficulties in mathematical analysis of intensional contexts, which in many cases are not compositional. This is true and also explains notorious difficulties encountered in formal analyses of modal propositions or epistemic reports. In fact, compositionality displays an essential feature of formal semantics. Languages and interpretations should be handled by mathematical devices operating according to recursive procedures. It requires that both correlate entities, i.e. languages and their interpretations, should have parallel properties. Additionally, compositionality allows the use of recursive and inductive procedures in syntax and semantics.

The introduced apparatus is not sufficient for interpreting sentences. It is an important point, because languages are set of sentences. We can interpret sentences in at least two different ways. Firstly, we can supplement a model **M** of **L** with two (I adopt the principle of bivalence) new distinct objects, namely **1** (truth) and **0** (falsehood) and assume that if a sentence *A* belongs to **L**, then **v**(*A*) = **1** or **v**(*A*) = **0**. Truth is not defined by this construction, but taken as a new primitive concept. This is Frege's way out, consistent with his view that logic is a language. Secondly, following Tarski, we can define the concept of truth in a well-known way, that is, as satisfaction of a sentence by all infinite sequences of objects; **1** and **0** can be thus dropped. An important difference between these two approaches, Fregean and Tarskian, consists in the fact that the former does not use a metalanguage while for the latter it is essential. Moreover, as Tarski explicitly showed on the occasion of forming the semantic truth-definition, the metalanguage cannot be reduced to the object language as far as the issue concerns semantic considerations. Consequently, if **L** is a language, and **ML** plays the role of its metalanguage, we have **L** ⊂ **ML**, but not (**L** = **ML**).

What is formal in performing interpretations of formalized languages? Of course, the answer depends on how the term 'formal' is understood. Suppose that something is formal if it is independent from content and solely dependent on form. This old meaning is fairly useful, but not very precise. As far as the matter concerns **L**, it is all simple, because the structure and signature of **L** are given in a purely syntactical way. The meaning (sense) of expressions of **L** is irrelevant here. Since we assume the signature parallelism between **L** and its models, the formal structure of the latter is displayed by the signature. Let us say that, for a given language **L**, its syntax and its signature display the form of interpretation given by the function **v**. On the other hand, **v** itself is not exclusively formal, because it depends on how expressions of **L** are understood. Let us take a language **L** with the alphabet $\{a_1, a_2, P_1, P_2\}$ Assume that: **U** is a set of philosophers, a_1 = 'Plato' a_2 = 'Aristotle', P_1 = 'a rationalist', P_2 = 'is older than'. The interpretation which is faithful to the history of philosophy correlates Plato with 'Plato', 'Aristotle' with Aristotle. Thus, the sentence 'Plato is a rationalist' is true, the sentence 'Aristotle is a rationalist' is false, the sentence 'Plato is older than Aristotle' is true and the sentence 'Aristotle is older than Plato' is false. Clearly, we can consider other interpretations in which things look differ ently, for example, possible worlds in which Plato was younger than Aristotle and Aristotle was a rationalist. However, regardless of whether or not our interpretations are faithful to empirical data and ordinary language, the function **v** behaves accordingly to some contents. Thus, we can say that the content of an interpretation is given through correlating **L** with some objects by the function **v**. Even if we say that this work is formal to some extent, because it is described by mathematical devices, like functions, sets, relations, etc., the meaning of the term 'formal' is different here. Anyway, the content of interpretation is obtained informally to some extent, contrary to the form of interpretation, which is defined inside syntax.

A general framework for the relation between the form and content of an interpretation can be outlined by appealing to one point in Tarski's construction of the truth-definition. It concerns the T-scheme, i.e. the equivalence *S* is true if and only if *S**, where the letter *S* denotes a name of the sentence in question and the symbol *S** denotes a metalinguistic translation of the sentence denoted by '*S*'. If the sentence denoted by '*S*' belong to **L,** the appropriate specification of (T) belongs to **ML**. The simple rule is as follows: if **L** is formal (formalized), **ML** is not. There is no other way to establish interpretation than to use more or less informal metalanguage. Thus, the content of interpretation goes from the top to the bottom, that is from **ML** to **L**. Beth (Beth 1962, p. XIV) once made the following remark:

> By *semantics* I mean a rigorously deductive treatment of the connections between the logical and mathematical symbols and the objects which they denote. An informal discussion of the same subject is denoted as *hermeneutics*.

This means that (a) hermeneutics consists in any decision how to interpret **L**, and (b) it is impossible to do formal semantics without a certain amount of hermeneutics or, equivalently, formal semantics is always embedded into some hermeneutics. Using the distinctions mentioned at the beginning of this paper, **LoC** and **LaC** are not sufficient for doing semantics of formalized. Even if we formalize **ML** in order to give a pure account of semantics of **L**, we must use an informal **MML**. This resembles the relation between formal and informal mathematics. We need the latter in order to develop the former.

Hintikka (see Hintikka 1988) is right to observe that the model-theoretic tradition considers calculi as re-interpretable systems, but it must be noted that every interpretation or re-interpretation requires a sufficiently powerful and partially informal metasystem. Needless to say, although logic as a calculus must be supplemented with logic as a language, the latter is the universal medium in Frege's or Wittgenstein's sense, at least if we assume the Tarskian way of doing semantics. Regardless of whether we work with the hierarchy of languages $\mathbf{L}_1$, $\mathbf{L}_2$ (= $\mathbf{ML}_1$), $\mathbf{L}_3$ (= $\mathbf{ML}_2$), ... (where $\mathbf{L}_n = \mathbf{ML}_{n-1}$) or claim that since everything can be captured by first-order languages, we need such languages and their extensions to higher-order ones, both hierarchies are open, provided that the last achieved level is partly informal. In light of metalogic, the Frege-Wittgenstein view that there are ultimate or universal borders of logic is mistaken. Although it is true, pace Wittgenstein, that logic itself does not exclude any possibility, this statement trivially follows from the metalogical theorem that pure logic does not distinguish any extralogical content. On the other hand, we can favour or exclude any possibility by performing a suitable interpretation in **ML**. Van Heijenoort says (Van Heijenoort 1967, p. 13) that Frege excludes any change of the universe. This is true, because the logical universe for Frege consists of the True and the False. It cannot be reduced to a single element due to a danger of a contradiction and cannot be expanded. On the other hand, the logical universe (the set of logical values) and the model-theoretic universe (the set of objects admissible as values of variables) are completely distinct objects and should be not confused. The present-day philosophy of logic can accommodate Frege's picture of the logical world without any difficulty. However, it has to reject his view concerning the universe of all possible objects. Finally, the universality of logic means that logical laws are universally valid, but it does not mean that logic sets the limits of extralogical discourse.

The distinction between **LoC** and **LoL**, which turned out to be useful in the history of logic, if it is taken as absolutely sharp, obscures the present logical scene. In particular, the limitative results of Gödel, Tarski and Church demonstrate limitations of syntax (calculus) and firmly support the view that semantics is prior to syntax or, in other words, that language is just prior to calculus. The very relation between syntax and semantics has some general consequences, which are very significant for the philosophy of logic and which cannot be expressed via the distinctions **LoC/LoL** and **LaC/LaL**. Let us suppose that one considers logic as the sum of syntax and semantics. It is clear that this sum is not homogeneous. Syntax or calculus are given effectively, which means that there is an effective procedure (algorithm) deciding in a finite number of steps whether a given syntactic construction is correct or not. Due to the parallelism of syntax and semantics effectivity also holds a property of the latter, but only partially. Some semantic properties, at least in classical logic, require infinitary proof methods. For example, the completeness theorem for first-order logic is not provable by finitary methods. Hence, the model-theoretic proofs of consistency of rich formal theories go beyond devices admissible for constructivists. Even if one formalizes semantics completely, this problem will automatically reoccur on the level of metasemantics. Suppose that one wants to justify a system of logic conceived as calculus, i.e. a deductive machinery by appealing to semantic properties. It is usually done by proving that some given logic is sound (provided that the premises are true, its rules lead to true conclusions) and complete (every tautology is provable). However, this justification must use methods going beyond the part of logic which is justified. Thus, logic as a syntactic and thereby effective construction, that is, formed according to recursive procedures, is justified by less effective constructions. At the same time, it is difficult to imagine a purely syntactic way of proving that some logic is sound and complete. It would be a paradox if logic was understood only as a calculus and such a proof would be impossible for logic conceived as the universal medium. This is the main reason why the distinction between syntax and semantics is more important than other distinctions indicated in the title of this paper. Finally, the Kantian metaphor that syntax without semantics is empty, but semantics without syntax is blind, means that the preciseness of calculus sharpens the semantic eye, although semantics brings the content into correct formulas.

III
Do We Need to Reform the Old T-Scheme?

Read 2010 almost verbatim reproduces Read 2008. I was invited to make comments about the later paper (see Woleński 2008). In what follows I shall partly repeat my earlier critical comments about Read's interpretation of the T-scheme, while also adding new remarks and improving the old ones; some of the additions are inspired by other contributors participating in Rahman, Tulenheimo, Genot 2008, including Stephen Read's replies in this volume (see Read 2008a) as well as the previous discussion in *Discusiones Filosóficas* (see Miller 2010, Read 2011, Sandu 2010).

At first, let me remind Leśniewski-Tarski's diagnosis of the Liar paradox. They pointed out (see Tarski 1944) that the derivation of the paradox uses: (I) self-referential sentences asserting semantic properties; (II) the T-scheme, and (III) classical, that is, bivalent logic. Hence, we can conceive three strategies in order to solve the paradox: (i) to exclude self-referentiality; (ii) to reject or modify the T-scheme; (iii) to change logic. It would be mistaken to maintain that there is a solution free of costs or some artificialities. It concerns Tarski who choose (i), Kripke (and many other logicians) who opted for (ii) (Sandu makes several remarks about this way out), and Read who tries to reform the T-scheme. More precisely, Read argues that the old (Tarskian) T-scheme is inaccurate and proposes its modification. I would like to show that Read's reading of Tarski is incorrect and that his (Read's) leads to some difficulties.

Read says that, according to Tarski, every instance of the T-scheme, that is the formula

(T) x is true if and only if p,

is true; Read even says that his understanding of (T) is "an unquestioned orthodox." However, Tarski focused on the provability of T-sentences from his truth-definition, but not on their truth. In Read's reply to my criticism (see Read 2008a, p. 218), he agrees with my standpoint, but he adds "But he [Tarski] clearly imposed that requirement because he thought those instances were true". I should note that Read uses the proper formulation in another place of his main paper (see Read 2010, p. 122). He probably thinks that both formulations are equivalent. However, they are not, unless we assume that the metatheory of truth-theory is ω-complete. Anyway, Read's statement about "an unquestioned orthodoxy" is certainly incorrect. For example, Miller and Sandu in their contributions published in *Discusiones Filosóficas* formulate (T) via provability, not truth.

I see no place in Tarski's writings which could justify the view that he "clearly imposed", etc. The observation that provable instances of (T) are true is trivial and has no relevance for Tarski's proposal of how to solve the Liar paradox. Nonetheless, the provability of T-sentences matters very much and the role of this fact has the best illustration in the problem of the definability of the T-predicate. Combining the fixed-point theorem and the Tarski undefinability theorem, leads to the unprovability of some instances of (T). This fact concerns the Liar sentence, independently of its formulation, as 'this sentence is false' or 'this sentence is not true', both recorded via arithmetization. The situation is clear in formal arithmetic of natural numbers, but the results that hold for arithmetic cannot be directly applied to natural language. However, pace Tarski, excluding the Liar sentence from the stock of permitted formulas of natural language plays a quite similar role to showing that some formalized instances of (T) are not provable. In fact, this is an analogical move as the exclusion of division by 0 in arithmetic. Although the formula $^{m}/_{n} = 0$ is grammatically correct and perfectly understandable, it must be rejected as producing inconsistency. As far as I know, Tarski never said that the Liar sentence is nonsensical, meaningless, etc. He only recommended that so-called closed languages, that is, languages containing a semantic concept used self-referentially, should be avoided. This restriction is by no means that counterexamples to (T) are excluded by *fiat*, as Read suggests (see Read 2010, p. 127). On the contrary, they are avoided by a subtle and elegant reasoning.

That T-sentences are assumed to be provable is important for their status. Read says (see Read 2010, p. 125) that according to Tarski (T) is "a merely a material equivalence." Consequently, the instances of (T) have the same status. Although Tarski himself was not quite explicit about this issue, it is rather obvious that provable theorems of the form $A \Leftrightarrow B$ are something more than material equivalences. In particular, one cannot replace A or B by their material equivalences. For instance, the equivalence: the sentence 'Stephen Read wrote a paper on T-scheme' is true if and only if Stephen Read wrote a paper about T-scheme' cannot be replaced by the equivalence: 'Stephen Read wrote a paper about T-scheme if and only if Jan Woleński commented on Read's paper', although both equivalences are true, and thus materially equivalent, and, moreover, both consist of true constituents. I do not suggest that Read considers such a replacement as possible or justified. Yet, I claim that the difference between merely material equivalences and provable material equivalences is important, particularly for a proper interpretation of Tarski's truth-theory.

Read entirely neglects some properties of languages for which the semantic concept of truth is defined. Firstly, such a language **L** is formalized or at least has a specified structure; this latter concept was introduced in Tarski 1944. Without entering into details, **L** must be well-described as a syntactic object. Firstly, we should know what belongs to the vocabulary of **L** and how the class of its sentences is defined. Secondly, and more importantly, **L** is an interpreted language. Let me quote the following words of Tarski (Tarski 1933, p. 166/167):

> It remains perhaps to add that we are not interested here in 'formal' languages and sciences in one special sense of the word 'formal', namely sciences to the signs and expressions of which no meaning is attached. For such sciences the problem here discussed has no relevance, it is not even meaningful. We shall always ascribe quite concrete and, for us, intelligible meanings to the signs which occur in the language we shall to consider. The expressions which we call sentences still remain sentences after the sign which occur in them have been translated into colloquial language. The sentences which are distinguished as axioms seem to us to be materially true, and in choosing rules of inferences we are always guided by the principle that when such rules are applied to true sentences the sentences obtained by their use should also be true.

Tarski did not explain what he understood by 'meaning'. However, it is clear that meanings of words, ordinary or artificial, for instance, the convention that 'black' expresses the property of being white, dictate semantic interpretations. Speaking more precisely, and using the model-theoretic terminology, meanings generate so-called interpretation functions which ascribe denotations of individual terms and predicates in given models. Although Tarski was skeptical as far as the matter concerns consistent semantic constructions for natural language, the difference between it and the interpreted, formalized or structurally specified **L** is less radical than it is frequently assumed by interpreters and critics of Tarski.

Taking into account certain facts about **L**, one can easily demonstrate that all of the counterexamples given by Read fail. This concerns the sentences:

(1) 'I am tired' is true if and only if I am tired;
(2) 'That book was stolen' is true if and only if that book was stolen;
(3) 'Any man is mortal' is true if and only if any man is mortal.

The argument is that the words 'I', 'that' and 'any' can have different meanings on the left and right sides of the respective equivalences. However, this cannot happen by definition in the case of properly interpreted languages. In particular, such languages do not require a special principle of uniformity (see Dutilh Novaes 2008), blocking ambiguities and other defects of expressions, because they are automatically excluded by technology of the semantic interpretation.

The valuation is a function and ‘I’, ‘that’ and ‘any’ have necessarily the same interpretations in all their occurrences in (1)–(3). This is particularly clear for indexicals, like ‘I’ and ‘that’. An ostensive specification of references (for example, supplementing the act of using ‘I’ or ‘that’ by a gesture) fixes the ascribed objects. On the other hand, if the references of such words are not made precise, they function as variables and the problem of truth (falsehood) of (1)–(3) does not arise at all. I do not argue that ambiguities do not occur in ordinary language, but only point out that the semantic definition of truth assumes that **L** for which it works is formalized (or has a specified structure) and its interpretation is fixed, even when **L** is selected as a suitable (not closed!) part of colloquial parlance. Read proposes to replace (T) by

(A) $\mathrm{T}x \Leftrightarrow \forall p(x{:}p \Rightarrow p)$.

This scheme codes an intuition expressed by

(S) *x* is true if and only if things are wholly as *x* says they are (or, however *x* says things are, they are).

Since what a sentence says is covered by all its implications, (A) formalizes (**S**), provided that the symbol $\Leftrightarrow$ denotes strict equivalence. Read says (Read 2010, p. 125) that “(A) is a logical equivalence” (expressed by the symbol $\Leftrightarrow$) contrary to (T) as a material equivalence. However, Read’s qualification of (A) as a the logical equivalence is vague. If we say that *A* is a logical sentence, we can mean various things. Firstly, (a) *A* is logical if and only if it is coded by signs belonging to the language of logic, but, secondly, (b) *A* is logical if and only if it is a logical theorem. Now, (T) and (A) are logical equivalences in the same sense, if the first understanding is assumed, but they both are not logical theorems. Thus, neither (a) nor (b) qualifies (A) as a logical equivalence, contrary to (T). Perhaps, Read intends to say that (A) is a logical equivalence, because strict connectives are logically stronger than material ones. However, due to my previous remarks about the status of the instances of (T), provable equivalences are “something more” than material equivalences. In fact, the difference between the formulas $X \vdash A \Leftrightarrow B$ and $A \Leftrightarrow B$ seems secondary, although, at least in my view, the former is much clearer than the latter.

Read points out that (A) is intensional for ‘says that’, but he does not see any problem with it. In particular, he seems to think that everything is solved by the closure of $x{:}p$ by “allowing substitution only of logical equivalents” (Read 2010, p. 124). Unfortunately, the matter is not so simple, because ‘says that’ is strongly intensional. Consider two equivalent formulations of a mathematical axiom, let say, \the parallel postulate. Denote them by *F* and *F’*, respectively. They are logically equivalent. Assume that we have a person *O* who does not know that *F* and *F'* are provably equivalent. Thus, the formulas x:*F* and x:*F'* do not say the same for *O*, although Read claims that they do. In fact, 'says that' forms contexts are not, contrary to Read, closed in being replaceable by their logical equivalents. Defining 'says that' as satisfying such a kind of closure is, in my opinion, at odds with the ordinary meaning of this operator.

There are other implausible consequences of (**A**). Consider (I do not assume that Peano arithmetic is first-order):

(4) T('Peano axioms') $\Leftrightarrow$ $\forall p$ ('Peano axioms': $p \rightarrow p$).

Peano axioms are true in both standard and non-standard models of arithmetic. Let **N** be the standard model and **N'** a nonstandard one. Take the sentence (*) 'all natural numbers have finitely many predecessors', which is true in **N**, but false in **N'**. Intuitively speaking, Peano axioms say (*) in **N**, but its negation in **N'**. Thus, 'says that' requires a relativisation to a model, but (T) without it has false instances. This also shows that defining the context x: p by the consequences of p may be insufficient in some cases, because a reference to models is required.

There is also a problem with (A) as applied to falsehoods (this is also pointed out by Miller and Sandu). The definition of F ('is false') corresponding to (A') can be recorded as

(A'') Fx $\Leftrightarrow$ $\exists p((x\colon p) \wedge \neg p)$.

As an example, we have

(5) F('Warsaw is the capital of France') $\Leftrightarrow$ $\exists p$('Warsaw is the capital of France': $p) \wedge \neg p$)).

However, (5) has its instantiation in

(6) F('Warsaw is the capital of France') $\Leftrightarrow$ (('Warsaw is the capital of France'): the greatest Polish city is the capital of France) $\wedge$ the greatest Polish city not is the capital of France.

Now assume that someone knows that a person *O* knows that the sentence 'the greatest Polish city is the capital of France' is false and that the right side of (6) is a correct (true) instantiation of the right side of (5). Hence, he or she knows that the sentence 'Warsaw is the capital of France' is also false; we use here the principle 'if an instantiation of *A* is false, *A* is false too'. However, to justify that, one must assume that 'Warsaw' and 'the greatest Polish city' are co-denotative (I neglect that the latter is a description). This consideration shows that we do not need to worry whether the sentences 'Warsaw is the capital of France' and 'the greatest Polish city is the capital of France' say the same or, eventually, in which circumstances they cover the same content, because it is sufficient to known the values of nominal expressions. Thus, even if we agree that that logically equivalent sentences say the same thing, this observation does not close the issue, because it can happen, as in the case of (5) and (6), that sentences are equivalent in theories or some language systems modulo denotative conventions, although they are not equivalent on purely logical grounds. Thus, the interpretation of a language has a crucial importance for establishing what sentences say and when they are true or false.

Further, the negation of (x: p) can be interpreted either as $\neg$(x: p) or as (x: $\neg p$) (the latter is stronger than the former). It matters in the case of negative sentences because

(7) T('Warsaw is not the capital of France') $\Leftrightarrow \forall p\,(\neg p \rightarrow$ 'Warsaw is not the capital of Poland': $\neg p$)

looks more plausible than

(8) T('Warsaw is not the capital of Poland') $\Leftrightarrow \forall p\,(\neg p \rightarrow$ (it is not the case that 'Warsaw is not the capital of Poland': $\neg p$)).

Perhaps the most important critical observation concerns metalogical properties of **T**. The law of the excluded middle can be stated as

(9) $\forall p(x{:}p \Rightarrow p) \vee \exists p((x{:}p) \wedge \neg p)$,

which is an instance of

(10) $\mathrm{T}A \vee \neg \mathrm{T}A$.

However, the relation of $\neg\mathrm{T}A$ and $\mathrm{T}\neg A$ is not clear under Read's definition. In general, we have

(11) $\mathrm{T}\neg A \rightarrow \neg \mathrm{T}A$,

and in the classical bivalent (two-valued) case the equivalences

(12) $\mathrm{T}\neg A \Leftrightarrow \neg \mathrm{T}A$, $\neg \mathrm{T}A \Leftrightarrow \mathrm{F}A$

hold. Now $\mathrm{T}\neg A$ has its interpretation (according to Read's definition) in

(13) $\forall p(x{:}\ \neg p \Rightarrow \neg p)$.

However, (13) does not imply

(14) $\exists p((x{:}\ p) \wedge \neg p)$,

although we have (or should have)

(15) $(x{:}\ \neg p) \rightarrow \neg(x{:}p)$.

I do not claim that (12) is indispensable, but only note that important relations are unclear in light of Read's proposals. In particular, the functor of negation commutes with truth (in its semantic understanding), but not with 'says that'.

Returning to '*x* says that *p*', a full analysis of this phrase seems to be much more complicated than Reads maintains. The general problem is that sentences can say quite different things for different persons or even for the same person depending on various pragmatic factors. For instance, we should distinguish direct (explicit) and indirect (implicit) 'saying that'. Every sentence entails infinitely many logical consequences. Even if we assume that the direct linguistic content of a sentence *A* is definable (it is a very optimistic presumption), the implicit content is much more vague and the amount of its grasping, always partial, cannot be accounted in advance. Reads seems to consider the intensionality of (A) as its advantage, but it is a very dubious view. I am inclined to think that the intensionality of (A) prevents a satisfactory definition of saying wholly as things are. This is very important, because the real virtue of the scheme introduced by Read essentially depends on such a definition.

(T) is essentially different from (A), because the former is purely extensional. In spite of this difference, I will argue that both schemes record almost the same intuitions. It is interesting that Tarski's starting point was similar to that proposed by Read. The initial intuition of the semantic definition of truth was presented by the following formulation (see Tarski 1933/1956, p. 155; Tarski followed Tadeusz Kotarbiński):

(*) A true sentence is one which says that the states of affairs is so and so, and the state of affairs indeed is so and so.

The original Polish formulation is much closer to (A), because it runs (in English translation): "a true sentence is a sentence which expresses that things are so and so, and things indeed are so and so". Tarski explicitly formalized (*) by T, presumably also for eliminating 'expresses' as an intensional factor. That **L** is interpreted and that its expressions have "intelligible meanings" illuminate the issue at stake. (T) can be supplemented by assertions of the type '*x* says that *p*', '*x* means that *p*', '*x* asserts that *p*', etc. All are external with respect to the instance of (T) and serve to fix, define, explain, etc. the valuation function connecting **L**-expressions with their denotations. However, '*x*:*p*', although internally embedded into (A), plays exactly the same role, provided that semantics a la Tarski is associated with this scheme. However, Read explicitly says (see Read 2010, p. 125) that this is just the case. In my opinion, (T) has similar intuitive advantages as (A), including the correspondence platitude, that is, the ability to express the basic content of *adequatio rei et intellectus.*

At the same time, the Tarski scheme avoids all problems caused by the internal intensionality of (A), in particular, the treatment of 'is false' (remember Russell's requirement that any satisfactory theory of truth should be also a theory of falsehood). In order to explain this point, let me refer to Miller 2010, p. 223). He attributes to Tarski the following scheme (I am using the notation of the present paper):

(T') $\mathrm{T}x \Leftrightarrow (x \in \mathbf{L}) \wedge p$.

The first conjunct in the right side of (T') indicates that the sentence named by x belongs to **L**. Miller define 'is false' by

(F') $\mathrm{F}x \Leftrightarrow (x \in \mathbf{L}) \wedge \neg p$.

Since Tx and Fx are mutual negations in classical logic, the same relation must hold between the right sides of (T') and (F'). However, $(x \in \mathbf{L}) \wedge p$ and $(x \in \mathbf{L}) \wedge \neg p$ do not negate each other. The negation of the former formula yields the sentence $\neg(x \in \mathbf{L}) \vee \neg p$. Assume that this disjunction is true in favour of $\neg(x \in \mathbf{L})$. Now, since x does not belong to **L**, it is not a sentence. Consequently, it is neither true nor false, because only sentences of **L** can be true or false. Clearly, Fx and ¬Tx are not equivalent under (T') and (F'). This example additionally shows difficulties of defining truth by conjunctions of conditions. It is much better to treat clauses, like $x \in \mathbf{L}$, as external with respect to (**T**).

Read claims that (A) is better, because it correctly solves the Liar paradox. According to Read, (T) is false, because it has a false instance (the letter l refers to the Liar sentence):

(#) l is true iff and only if l is not true.

Still, (A) produces a correct truth-condition of l in the form:

(##) $\mathrm{T}l \Leftrightarrow \neg \mathrm{T}l \wedge \mathrm{T}l$.

However. this success is entirely apparent. The formula (#) does not give a complete analysis of the Liar sentence. First of all, (#), as inconsistent, is not provable in the semantic theory of truth. Since (##) is true for Read, the formula Tl is either false or inconsistent. Yet, both these cases must be excluded. A simple transformation of (##) gives:

(####) $\neg \mathrm{T}l \Leftrightarrow \mathrm{T}l \vee \neg \mathrm{T}l$,

which is not satisfactory, because it says that a tautology is not true, that is, inconsistent. In fact, adding ¬Tl (this is an unprovable formula!) to the stock of theorems of propositional logic immediately abolishes the Post (absolute) consistency of this logic. The actual situation of l is displayed by the formula

(####) $Tl \Leftrightarrow Fl \Leftrightarrow \neg Tl \wedge Tl$.

Roughly speaking, every semantic assumption about *l*, that is, every assumption concerning its truth-value modulo the bivalence, entails a contradiction. It clearly shows, firstly, why some instances of (T) are excluded as unprovable and, secondly, that (##) does not formulate any truth-condition. This concurs with Sandu's remark that (see Sandu 2010, p. 289) it is not clear what the Liar sentence says, although I am inclined to a more radical conclusion, namely that *l* says nothing. To sum up, if (A) is reduced to (T), the former solves the Liar paradox in the same way as the latter does, but if is not reduced, the issue is still open, because the proponent of (A) must decide whether this schema has true, but unprovable instances.

Finally, I would like to note that the problem of a philosophical significance of the T-scheme is still open. Does it code the correspondence intuition or not? Should we modify (**T**) in order to express the correspondence platitude? I do not address these and similar questions in this paper. I hope to give an account of the philosophical significance, if any, of Tarski's theory of truth in my Woleński (in preparation.

IV
Truth is Eternal if and only if It is Sempiternal

The problem addressed in this paper goes back to Aristotle and his considerations about tomorrow's sea battle. In a famous passage in *De Interpretatione* (19a 25–30; after *The Works of Aristotle*, vol. 1: *Categoriae and De Intepretatione*, tr. by E. M. Edghill, Oxford University Press, Oxford 1928), the Stagirite says:

> Everything must be either be or not be, whether in the present or in the future, but it is not always possible to distinguish and state determinately which of these alternatives must necessarily come about.
> Let me illustrate. A sea-fight must take place to-morrow or not, but it is not necessary that it either should not take place to-morrow, neither it is necessary that it should not take place, yet it is necessary that it either should or should not take place to-morrow. Since propositions correspond with facts, it is evident that when in future events there is a real alternative, and a potentiality in contrary directions, the corresponding affirmation and denial have the same character.

These words initiated a considerable discussion about the relation between truth and time. Is truth relative and dependent on temporal coordinates, or is it absolute and timeless? The debate concerns several problems, in particular, the validity of some logical principles, fatalism, God's omniscience, free-will and determinism (see Bernstein 1992, Cahn 1967, Gaskin 1995, Hintikka 1977, Lucas 1989, Prior 1953, Vuellimin 1996). This paper concentrates almost entirely on logical issues.

Interpretations of Aristotle's quoted passage essentially vary. Did he revise logic or not? Literally understood, his text defends the principle of excluded middle (**EM** for brevity), but it also tells us something about the logical value of future contingents, that is, sentences about accidental events which will happen or not. Should we distinguish the predicates 'true' and 'true at a time-moment *t*'? This problem was also noted by the Stoics. Their celebrated Master Argument tried to prove that everything that is possible will happen in the closer or further future. Thus, **EM** is valid and truth has no temporal coordinates. Jan Łukasiewicz (see Łukasiewicz 1930) interpreted the Stoics as defending the principle of bivalence (**BI**). According to him, we should distinguish the logical and metalogical versions of **EM** (**MEM**). Whereas the former is a logical theorem coded by the formula $A \vee \neg A$ (or its another notational variant), the latter says that every sentence is either true or false. **BI** conjoins **MEM** with the metalogical law of non-contradiction (**MNC**), that is, the assertion that no sentence is both true and false. Łukasiewicz interpreted Aristotle as suggesting that future contingents are neither true nor false. Thus, in light of Łukasiewicz's interpretation, Aristotle limited **BI**.

Łukasiewicz's own solution followed his reading of the Stagirite. He rejected **BI** as universally valid and envisaged three-valued logic (later, many-valued, countable or even uncountable, systems). According to him (I restrict the discussion to three-valued logic, that is, $\mathbf{Ł}_3$), if *A* is a sentence about a future contingent event, *A* possesses the third logical value; let us call it neutrum (Łukasiewicz's first idea consisted in considering neutrum with possibility). **EM** is not a logical theorem. Numerically speaking, if *A* possesses neutrum as its logical value, $v(A) = ½$. Furthermore, $v(A) = ½ \Leftrightarrow v(\neg A) = ½$. The valuation for disjunction in $\mathbf{Ł}_3$ establishes that if $v(A) = v(B) = ½$, $v(A \vee B) = ½$ as well. Hence, **EM** is not a theorem of $\mathbf{Ł}_3$, because *A* is a theorem of this logic if and only if $v(A) = 1$, for any valuations of components of *A*. The logical law of non-contradiction, that is, the formula $\neg(A \wedge \neg A)$ (**NC**) shares the fate of **EM** in $\mathbf{Ł}_3$, because if $v(A) = v(B) = ½$, $v(A \wedge B) = ½$. Yet $\mathbf{Ł}_3$, as any other logical system, has to satisfy the condition of consistency. Thus, **MNC** holds in the metatheory of $\mathbf{Ł}_3$. This is a very strong reason to distinguish **NC** and **NMC**. However, Łukasiewicz still considered truth as absolute. According to him (I restrict the discussion to three-valued logic, that is, $\mathbf{Ł}_3$), if *A* is a sentence about a future contingent event, *A* possesses the third logical value; let call it neutrum (Łukasiewicz's first idea consisted in considering neutrum with possibility). It will true or false in the future. Truth and falsity are eternal, but not sempiternal. This means that truth and falsity cannot change in their opposite or other logical values. On the other hand, sentences valued by the neutrum can become true or false; Łukasiewicz did not consider such changes, deeming logical values as necessary. This means that, at least for Łukasiewicz, future contingents can become true or false, but it is not necessary. Łukasiewicz's view has important and deep consequences for understanding the absoluteness of truth. His position was moderate, because eternality suffices for absoluteness.

The equivalence 'truth is absolute if and only if truth is eternal' can be called the weak truth-absoluteness thesis (**WTAT**). The strong truth-absoluteness thesis (**STAT**) says that truth is absolute if and only if truth is eternal and sempiternal. **STAT** can be stated as

(1) If a sentence is true at a moment *t*, it is also true at any other moment *t'*.

This statement has two parts:

(1a) (eternality, **ET**) If a sentence is true at *t*, it is also true at any $t' \geq t$;

(1b) (sempiternality, **ST**) If a sentence is true at t, it is also true at any $t' \leq t$;

Consider a concrete example, namely the sentence (*) I will go to Warsaw on April 26, 2012; the date of writing this sentence is April 26, 2011. Assume that I will stay in Warsaw on April 2012. Thus, (*) is true on t = April 26, 2012. Consequently, we have:

(2a) (**ET**$^{(*)}$) If (*) is true at t, it is also true at any $t_1 \geq t$;

(2b) (**ST**$^{(*)}$) If (*) is true at t = April 26, 2012, it is also true at any $t_1 \leq t$.

Clearly, if A is valued by the neutrum, being an indefinite sentence, $v(A)$ = ½), (1a) holds, but (1b) loses its valour.

STAT was defended in Poland by Kazimierz Twardowski (see Twardowski 1902) and Stanisław Leśniewski (see Leśniewski 1912) even before Łukasiewicz envisaged many-valued logic. Twardowski used philosophical (epistemological) arguments. He criticized all forms of truth-relativism, not only the view that truth is essentially determined by temporal coordinates (see Kokoszyńska 1948, Kokoszyńska 1951 for a modern shaping of Twardowski's arguments). As far as the matter concerns logic, Twardowski observed that the relativity of truth is at odds with the principle non-contradiction (he did not distinguish its logical and metalogical formulations; the same concerns Leśniewski in his early writings). Leśniewski offered a proof that, assuming the law of non-contradiction, **ET** and **ST** are equivalent, that is, **STAT** holds. His argument is as follows. Assume that a sentence (**) 'S is P' is true, but not sempiternally true. Thus, there is a time t in which (**) is not true, although it is true at some $t_1 \geq t$. This assertion implies that (***) 'S is not P' is true at t. Leśniewski uses the principle of non-contradiction at this point and concludes that since (***) is always false, it is also not true at t. Thereby, (**) must be true at t, contrary to our initial assumption. However, this produces inconsistency. Leśniewski's proof that eternality entails sempiternality proceeds analogously.

Leśniewski's reasoning is informal. One of its steps can be questioned, namely the one in which Leśniewski assumes that falsehood is sempiternal. One might observe that if we assume that the property of 'being false' obtains sempiternally, the same concerns truth automatically. Fortunately, the outlined proof has its contemporary setting, which does not require any appeal to the sempiternality of falsehood. Assume that (a) **X** is the true set of sentences about the past; (b) A is a future contingent; (c) A is independent of **X**. The supposition (c) is necessary, because the dependency of A would imply **STAT**. The assumption (a) implies that **X** is consistent and thereby possesses a model. Due to (c), the sets $\mathbf{X}_1 = \mathbf{X} \cup \{A\}$ and $\mathbf{X}_2 \cup \{\neg A\}$ are consistent as well and have models, let us call them $\mathbf{M}_1$ and $\mathbf{M}_2$. We can think about $\mathbf{M}_1$ and $\mathbf{M}_2$ as models of maximally consistent sets of sentences; this is perhaps the best intuition related to the concept of possible worlds. In virtue of Lindembaum's theorem on maximalization, every consistent set of sentences has its maximal extension. Thus, the set **X** has at least two such extensions which contain the sets $\mathbf{X}_1$ and $\mathbf{X}_2$. Similarly, both models $\mathbf{M}_1$ and $\mathbf{M}_2$ extend the model **M**. More technically, $\mathbf{M}_1$ and $\mathbf{M}_2$ are submodels of **M**. Clearly, sentences true in **M** remain true in $\mathbf{M}_1$ and $\mathbf{M}_2$.

Another construction which justifies (c) is as follows. We define (see Asser 1972, pp. 168–169) the concept of branchability by:

(3) (a) a set **Y** of formulas branches at a formula A iff the sets $\mathbf{Y} \cup \{A\}$ and $\mathbf{Y} \cup \{\neg A\}$ are consistent;
(b) a set **Y** is branchable iff there is a formula A at which X branches;
(c) a set **Y** is branchable iff **Y** is a consistent and incomplete set of sentences.

Since we have no reason to assume that **X** is a complete set of sentences, we say that **X** branches at A. This is displayed by the diagram (**Δ**).

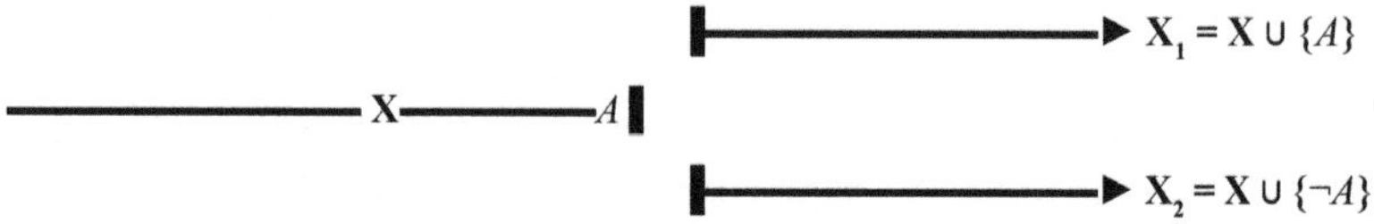

An ontological interpretation is also possible. Let A be a sentence uttered at the moment fixed as present (denoted by t) and referring to a future event **E,** which will happen at t_1 ($\mathbf{E}(t_1)$ or will not happen at t_1 ($-\mathbf{E}(t_1)$)); the segment indicated by the **PAST** covers everything what happened before t (including this moment). We transform the diagram (**Δ**) into (**Δ'**)

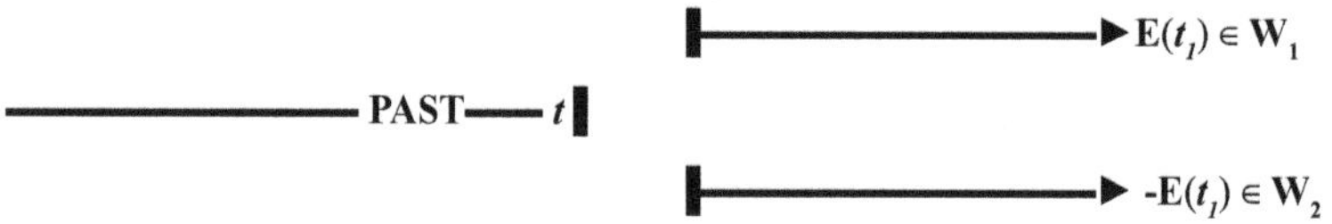

$\mathbf{W}_1$ and $\mathbf{W}_2$ are possible worlds, that is, possible continuations of **PAST** as their initial segment. Otherwise speaking, $\mathbf{W}_1$ and $\mathbf{W}_2$ enlarge **PAST**, but in radically different (or ontologically inconsistent) ways, because the former validates A and the latter verifies $\neg A$.

The status of **M** as a submodel of $\mathbf{M}_1$ and $\mathbf{M}_2$ justifies **ET**. In fact, if A will be true at the point t_1 later than t, its truth becomes stable for ever. Thus, **ET** is automatically valid. For the sake of the further argument, we can consider the value of A at any temporal point t earlier that at t_1, not necessarily at the moment in which this sentence is (was) issued. Assume that A is eternally true, that is, if A is true at t, it remains true at any $t_1 \geq t$. This means that A is true in $\mathbf{M}_1$. However, the sentence in question cannot be false in any submodel of $\mathbf{M}_1$, in particular in **M**. Hence, we immediately conclude that if A is eternally true, it is sempiternally true as well. The same line of reasoning applies to the assumption that $\neg A$ as true in $\mathbf{M}_2$, implies that if A is false in $\mathbf{M}_2$, it is also false in **M**. Thus, both logical values are stable over time, **ET** and **ST** are equivalent, and **STAT** finds its support. Now, we have exact tools for comparing many-valueness, **BI** and truth-absoluteness. The argument for **STAT** essentially uses **BI**. The converse dependence holds as well. Hence, **BI** and **STAT** are equivalent, at least in the model-theoretic semantics. Nonetheless, if the truth-absoluteness is reduced to eternality, **BI** has no explicit connection with **ET**. One can assume that any indefinite sentence becomes true or false sooner or later, but the supposition that at least some indefinite sentences remain such for eternity is also possible. In both cases, however, **ET** can be defended. Thus, Łukasieiwicz could consider truth as absolute, but provided that this property would be understood in a weaker sense, that is, according to **WTAT**.

Many philosophers, and Łukasiewicz is a good example in this respect, accept the eternality of truth as uncontroversial, but complain about sempiternality. The argument for the logical equivalence of (**ET**) and (**ST**) shows that something is wrong in this view, because if sempiternality is felt as non-intuitive, the same should concern eternality. I only indicate this problem briefly. Its full analysis requires taking in account ontological questions related to determinism and its various forms. For example, radical determinism can be displayed by the diagram (**Λ**")

——————— PAST—— t ———— FUTURE ———►

This line symbolizes the uniform "flow" of reality from the past though the present moment t to the future. This flow can be organized by the strict (one-to-one) causal connection. In fact, Łukasiewicz argued that sempiternality is implied by **MEM** plus causality. However, the outlined argument shows that **STAT** can be defended on purely logical grounds and without an appeal to ontology. Thus, objections against sempiternality as an intuitive property of truth seem to come as a result of a conflation of logic with ontology.

Let me add that that my reasoning concerns truth and falsehood in their model-theoretic sense. Perhaps, Leśniewski's demonstration is semantic in principle. Anyway, his argument implicitly uses bivalence, namely by employing the reductio ad absurdum mode. Thus, the conjunction of the reduction principle and the principle of non-contradiction has the same metalogical consequences as the product of **MEM** and **MNC.** Anyway, classical logic is closely related to the absolute treatment of both logical values. Moreover, the semantic definition of truth is an example, which suggests an absolutists interpretation. In fact, some authors (see Kokoszyńska 1936, Woleński 1994) argue that Tarski defined the absolute concept of truth.[2] Leśniewski (see Leśniewski 1929) suggested – and this remark applies to the semantic concept of truth – that we should not use the predicate 'is true at *t*', but rather the construction 'is *P* at *t*' is true (false)'. In other words, truth has no temporal coordinates, but predication has or might have them. But different solutions are also possible. Borkowski (see Borkowski 1991) claims that Łukasiewicz simultaneously worked with two different semantic valuations represented by the pairs {true, false} and {true at *t*, false at *t*). Suszko (see Suszko 1975) distinguished the logical valuation (true/false) and algebraic valuations consisting in introducing various objects (in particular, situations) denoted by sentences. These constructions provide other tools for defending **STAT** as related to the basic valuation of sentences by the unqualified truth and the unqualified falsehood. All mentioned strategies are explicitly model-theoretic or convertible to such. Let me, however, remark that there is a question of how the relativity/absoluteness debate over the concept of truth could be applied to correspondence theories, for example, Hochberg's explication (see Hochberg 1984a) of the relation between fact and truth, particularly if the concept of truth-makers is introduced. I see some problems here associated with the absence and presence of facts, because truth-makers, considered as facts, states of affairs, situations or other objective (ontological) correlates of sentences, seem to require additional valuations.

2 What is more, Kokoszyńska identified the classical definition of truth with the absolutist account of this notion (see Kokoszyńska 1936).

V
Is Identity a Logical Constant and are there Accidental Identities?[3]

Propositional connectives and quantifiers are logical constants without any doubt. However, we speak about first-order logic with or without identity. Even this way of speaking suggests that identity has a special status to some extent. In fact, the status of identity is controversial. Wittgenstein says (Wittgenstein 1922, 5.5303):

> Roughly speaking, to say of *two* things that they are identical is nonsense, and to say of *one* thing that it is identical with itself is to say nothing at all.

According to Wittgenstein, identity is not a relation. This view rises an important point: does identity hold between objects, which are numerically different, for example, two occurrences of 'e' in the word 'different'? Tarski's view of identity was radically opposite to that of Wittgenstein. The difference is well illustrated by the following quotation (Tarski 1941, p. 49):

> Among the logical concepts not belonging to sentential calculus, the concept of IDENTITY, or, of EQUALITY, is perhaps the one which has the greatest importance.

Wittgenstein's and Tarski's statements about identity can be rephrased without essential changes by replacing 'identity as a relation' by 'identity as a predicate' (I prefer the second way of speaking).

Formally speaking, identity is introduced into first-order logic by the axioms (I omit quantifiers in the front of formulas):

(A1) $x = x$;
(A2) $(x = y) \Leftrightarrow (y = x)$;
(A3) $(x = y) \land (y = z) \Leftrightarrow (x = z)$,

together with the rule of replacement (for simplicity, I restrict it to monadic predicates, but its is obvious how to generalize this rule from arbitrary formulas), which can be formulated as:

(RR) $(x = y) \Leftrightarrow (Px \Leftrightarrow Px/y)$.

3 This paper uses and extends some considerations in Woleński 2005 and Woleński 2008.

Thus, first-order logic with identity (**FOLI**) is determined by the propositional calculus (**PC**; more precisely, by its codification via axioms or rules of inference). Pure (the meaning of 'pure' in this context will be explained below) first-order logic without identity (**FOL**) is codified by axiomatization or rules of inference and the set {(A1)–(A3), (RR)}.

As it is well known, the identity predicate is not definable in first-order logic. The situation changes in second-order logic via the Leibniz rule:

(LR) $(x = y) \Leftrightarrow \forall P(P\mathrm{x} \Leftrightarrow Py)$,

which says that identical objects have the same properties. In fact, the implication

(1) $(x = y) \Rightarrow \forall P(P\mathrm{x} \Leftrightarrow Py)$

suffices for defining identity. The reverse implication

(2) $\forall P(P\mathrm{x} \Leftrightarrow Py) \Rightarrow (x = y)$,

expresses the famous *principium identitatis indiscernibilium* (the principle of identity of indiscernibles). Hence, (LR) is the conjunction of (1) (the principle of indiscernibility of identicals) and *principium identitatis indiscernibilium.*

Although formal properties of identity are (or seem to be) clear, the concept of identity provides several problems for logicians and philosophers. One of them is captured by the already mentioned question: 'Is identity a logical constant?' The arguments for the affirmative answer point out that fundamental metalogical results (semantic completeness, compactness, undecidability, the Löwenheim-Skolem theorem, the Lindström theorem) are valid for **FOLI**. In particular, the last result seems important because it provided a characterization of first-order logic as contrasted with higher-order logic. Consequently, the the Lindström theorem determines the borderline between 'being the logical' and 'being the extralogical', that is, the first-order thesis (see Woleński 2004a for details and a discussion). Speaking metalogically, all theorems with the identity-predicate derivable in **FOLI** are universally valid. Speaking more philosophically, these theorems are necessary in the strongest sense, because logic represents an uncontroversial kind of necessity. Tarski (see 1986a) argued that identity is a logical notion because it is invariant under all transformations of a domain into itself.

However, there are some problems with considering identity as a purely logical item. Having identity, we can define numerical quantifiers of the type 'there are *n* objects', where *n* is an arbitrary natural number. Consequently, we can characterize finite domains, although first-order logic is too weak to define the concept of finiteness. Now, if we add the sentence 'there are n objects' to first-order logic, its theorems are valid not universally, but in domains that have exactly n elements. Hence, it seems that identity brings some extralogical content to pure logic, contrary to the view (it can be expressed by a suitable metalogical theorem) that logic does not distinguish any extralogical content. Perhaps, this is the very reason why the label 'the identity-predicate' is used, although logicians simultaneously remark that this is a very special predicate. Anyway, a qualification of identity as logical or extra-logical is conventional to some extent.

Another reason to see identity as an extralogical problem stems from the so-called inflation and deflation theorems (see Massey 1970), both closely related to the definability of finite domains in **FOLI**. The former says that if a formula, say A, is satisfied in a non-empty domain **D** with n elements, it is also satisfied in any domain **D'** with at least n elements. The deflation theorem asserts that if A is satisfied in **D**, it is also satisfied in any **D'** with at most n elements. Although these theorems hold for **FOL**, they fail for **FOLI**. The formula $xy(x = y)$ provides a counterexample to the inflation theorem, because it is true in the one-element domain and no other, but the formula $xy(x\ ^1\ y)$ is false in the domain with one element. On the one hand, both theorems hold in **FOL**. This is a reason for applying the adjective 'pure' to first-order logic without identity. On the other hand, if we look at **PC** and **FOL**, we can note some metalogical difference between both systems. In particular, **PC** is decidable, but **FOL** has no decision procedure. Furthermore, **PC** is Post-complete, but **FOL** (with numerical quantifiers) lacks this property. This shows that the concept denoted by the phrase 'being the logical' has a different strength in particular subsystems of **FOL**.

One additional problem requires a clarification. According to early Frege (see Frege 1879) and Wittgenstein (see Wittgenstein 1922), identity operates on signs. The view that the formula $x = y$ concerns objects became standard in contemporary logic. However, the notation used in (A1) – (A3) (as well as in other quoted formulas) is ambiguous to some extent. In fact, under the objectual treatment of identity, we should formulate (RR) as

(RR') if $d(x) = d(y)$, then $P(x) \Leftrightarrow \mathrm{P}(x/y)$.

This formula means: if the object denoted by the letter x (the denotation of x) is identical with the object denoted by the letter y (denotation of y), then if the denotation of x has a property P, the letter x can be replaced by the letter y. Note that the antecedent of (RR') concerns objects, but its consequent deals with objects and signs. A suitable rephrasing of other listed formulas is straightforward. The proposed reading of the replacement rule underlines its semantic character. We should think about identity as determined by an interpretation of terms in models; denotations depend on valuation functions ascribing objects to terms (individual constants and individual variables; I omit valuations of predicated letters). Note that (A1) is the only axiom, which is an unconditional formula contrary to (A2), (A3) and (RR) [or (RR')]. I assume the objectual framing of identity in what follows without using the symbolism employed in (RR').

Kripke (see Kripke 1971) presents an argument intended to show that there are no accidental identities or that every identity is necessary. This view is supported by the following reasoning. Assume (RR) and (A1) in the following form (I insert quantifiers, because their position plays a relevant role in the argument; the box expresses necessity):

(A1') $\forall x\Box(x = x)$ (every objects is necessarily self-identical).

Now, interpret P as the property 'necessarily identical with'. (RR) gives

(3) $\forall xy(x = y) \Leftrightarrow (\Box(x = x) \Rightarrow \Box(x = y))$.

Since $\Box(x = x)$ is universally valid, it can be omitted. Thus, we obtain:

(4) $\forall xy(x = y) \Rightarrow \Box(x = y))$,

that is, the conclusion that if two objects are identical, they are necessarily identical. However, this result seems non-intuitive, because the identity of London and the capital of UK looks as accidental.

How convincing is this reasoning? First of all, let us change it by using the provability operator $\vdash$. Since (A1) – (A3) are logical axioms (I assume here that identity is a logical constant), we can add $\vdash$. As far as the matter concerns (RR) (translated into a formula), we obtain (quantifiers added):

(RR'') $\forall xy(\vdash(x = y) \Rightarrow (Px \Rightarrow Py))$.

Since the provability operator is monotonic, (RR'') entails

(5) $\forall xy(\vdash(x = y) \Rightarrow \vdash(Px \Rightarrow Py))$.

Two things are to observed. Firstly, the antecedent inside (5) has the sign $\vdash$. Is it possible to skip this element in (5). It would be at odds with the practice of using identity in inferences. For instance, mathematicians derive conclusions about properties of identical objects, assuming that their identity is provable in mathematical theories. Secondly, we cannot interpret P as expressing provability. Now, if provability is understood as a kind of necessity, Kripke's argument cannot by repeated. We can only obtain:

(6) $\forall xy(\Box(x = y) \Rightarrow \vdash\Box(Px \Rightarrow Py))$.

Inserting or omitting the formula $\Box(x = x)$ is completely pointless in this reasoning. Let us strengthen (5) to the formula:

(7) $\forall xy(\vdash (x = y) \Leftrightarrow \vdash (Px \Leftrightarrow Py))$.

We can think about (7) as a scheme capturing the first-order version of the Leibniz rule. Disregarding whether the provability of the right part of (7) is realistic or not, this equivalence leads to:

(8) $\forall xy(\Box (x = y) \Leftrightarrow \Box (x = y))$,

which is not very exciting, because it asserts that if an identity is necessary, it is necessary.

I will not discuss Kripke's solution of the puzzle produced by (5) in details (he accepts that some identities are accidental and a posteriori). I only note that his view assumes few things, in particular, the distinction between rigid and non-rigid designators and essentialism as well as admissibility of switching from *de dicto* necessities to *de re* ones. There is no essential difference between (A1) and (A1'), although the latter has the *de re* form (the quantifiers precede the box). However, replacing *P* by 'necessarily identical with' relevantly uses the *de re* formulation. My proposal consciously ignores all exrtalogical circumstances except the claim that necessity should be used *de dicto* and as related to the provability. This blocks the passing from $\Box(x = y)$ to $x = y$. Without assuming $x = y$ as independent of $\Box$ or $\vdash$, the conclusion that all identities are necessary does not follow. And yet we can still keep the difference between the unconditional, like (A1), identity validities and the conditional ones, namely (A2), (A3). Facts of interpretation are, of course, accidental and a posteriori; for instance, the term 'the capital of UK' could be valued not by London, but by another British city, say, Manchester. However, if an interpretation is fixed, its consequences are conditionally necessary. Since unconditional identity is a special case of conditional, we obtain a uniform treatment of =, independent of the view whether it is a logical constant or not. These conclusions are obvious, if we adopt the objectual understanding of identity.

VI
Naturalism and Genesis of Logic

It is traditionally accepted that we differentiate between *logica docens* and *logica utens*, that is, between theoretical logic (logic as theory) and applied or practical logic. Both can be defined with the use of the concept of logical consequence. The first is a set of consequences of an empty set, symbolically **LOGT** = $Cn\varnothing$, provided that the operation Cn satisfies the well-known general Tarski's axioms, i.e. denumerability of the language (a set of sentences) **L**, $X \subseteq CnX$ (the inclusion axiom; **X**, **Y** are sets of sentences of **L**), if $X \subseteq Y$, then $Cn\,\mathbf{X} \subseteq \mathrm{Cn}\mathbf{Y}$ (monotonicity of Cn), $CnCn\mathbf{X} = Cn$ (idempotence of Cn) and, if $A \in Cn\mathbf{X}$, then there is a finite set $\mathbf{Y} \subseteq \mathbf{X}$ such that $A \in Cn\mathbf{Y}$ (Cn is finitary). Cn is a mapping of the type $2^{\mathbf{L}} \rightarrow 2^{\mathbf{L}}$, that is, transforming subsets of **L** into its subsets. In order to make things simpler, I assume that Cn is based on classical logic. **LOGT** can be also defined as the only common part of the consequences of all sets of sentences. Otherwise speaking, logic is the only non-empty intersection of the family of all subsets of **L**. What follows from it is that logic is included in the consequences of each set of sentences, which underlines its universal character. If $Cn\mathbf{X} \subseteq \mathbf{X}$, then $X = Cn\mathbf{X}$ due to the inclusion axiom. Moreover, if $Cn\mathbf{X} \subseteq \mathbf{X}$, we say that that **X** is closed by the consequence operation of logical consequence. This is the definition of a deductive system (a deductive theory). Thus, in the case of deductive systems, Cn does not extend **X** beyond itself. The concept of logical consequence belongs to the syntax of language. The notion of logical following (entailment) is a semantic counterpart of Cn. The properties of both of these notions are such that if $A \in Cn\mathbf{X}$ and X consists of true sentences, the sentence A must be also true. If **X** is a theory and the set $Cn\mathbf{X}$ coincides with a set of true sentences (specifically: true in a determined model **M** or relevant class of models) of **X**, this theory is semantically complete.

The statement that Cn closes sets of sentences as long as $Cn\,\mathbf{X} \subseteq \mathbf{X}$ suggests some analogies with topology, since certain properties of this operation satisfy Kuratowski's axioms for topological spaces. Let Cl denote the closure operation of a topological space, and **X**, **Y** – any subspaces (subsets); I intentionally use the same letters for denoting the set of sentences and sets investigated by topology. Then (see Duda 1986, p. 115): (1) $Cl\varnothing = \varnothing$; (2) (2) $X \subseteq ClX$, (3) $ClCl\mathbf{X} = Cl\mathbf{X}$; (4) $Cl(\mathbf{X} \cup \mathbf{Y}) = Cl\mathbf{X} \cup Cl\mathbf{Y}$. Operations Cn and Cl differ from each other as far as the matter concerns axioms (1) and (3), because, in the case of logic, the set $Cn\varnothing$ is non-empty and $Cn\mathbf{X} \cup Cn\mathbf{Y} \subseteq Cn\mathbf{Y}$ but the reverse inclusion does not hold. The first difference is founded on the specific definition of logic, which does not possess a clear topological sense (I will return to this question below), while the other one indicates a partial analogy between closed sets in the topological sense and deductive systems in the logical sense, because (3) does not hold for arbitrary sets of sentences. Thus, "logical" closure is weaker than the topological one. The set of theses of logic is for sure non-empty and it is a system. It can be treated as a specifically closed topological space, with individual theorems as its points.

Topology$\{\varnothing = Cl\varnothing, \mathbf{X}\}$ is minimal (see Wereński 2007, p. 124) in the sense that the smallest one cannot be examined. Next, $Cl\varnothing \subseteq ClX$, since for each **X**, $\varnothing \subseteq X$. Let us agree (this is a convention) that $Cl\varnothing$ is a topological equivalent of logic. The motivation for this convention consists in taking into consideration that a proof of logical theorems does not require any assumption. The evident artificiality of this convention can be essentially weakened by the acknowledgment that closing an empty set produces any theorem of logic. It can be shown that if A and B are theses of logic, then $Cn\{A\} = Cn\{B\}$, which means that any two logical truths are deductively equivalent. Let us assume that X (this time as a set of sentences) is consistent and consists of the set **X'** of logical tautologies and a set **X"** of theorems outside logic. Thus, $\mathbf{X'} = Cn\varnothing$ and $\mathbf{X''} \subseteq \mathbf{J} - \mathbf{X'}$. Sets **X'** and **X"** are disjoint and constitute mutual complements in the set (space) **X**. Since the set **X'** is closed, its complement, i.e. **X"**, is an open set. The introduced convention about $Cl\varnothing$ allows to "topologize" the properties of sets of theses; in particular, it makes it possible to treat the set **X** (of theses) as a clopen set. From the intuitive point of view, the operation of logical consequence encodes inference rules of deriving some sentences from other sentences, that is, the procedure of deduction of conclusions from defined sets of premises. Deduction, at the same time, is infallible, that is, it never leads from truth to falsity.

What is applied logic or *logica utens*? When **X** is any non-empty set of sentences, then applied logic **LOGA(X)** of this set can be associated with the operation *Cn* applied to **X**. This is applied logic in a potential sense. This understanding of *logica utens* is, however, decidedly unrealistic, since its user applies only these rules that he needs, regardless of whether or not he does so in a conscious way. In other words, real applied logic of a given set is a stock of these logical laws (or rules) that are used in a concrete inferential work. This circumstance makes it impossible to give an abstract definition of real applied logic. It is worth to observe that *Cn* can be based on a non-classic logic, e.g. intuitionistic, many-valued or modal logic. Furthermore, we can neglect the monotonicity condition in order to obtain a non-monotonic logic. These remarks point to the fact that non-classical logics is similarly definable as the classical system. Since applied logic operates on closed-open sets, they, by this assumption, contain extra-logical sentences beside theorems of logic. The inclusion condition decides that logic can be deduced from any set of sentences. This fact has a serious methodological importance. If deduction within closed sets 'leads' to accumulation points in the topological sense, adding new extralogical sentences can be executed in an extra-deductive manner. This corresponds to the definition of an open set as such that includes all of its neighbourhoods. To put it in a different way, the transition to neighbourhoods of sentences as points in spaces in the set **X**", that is, the extension of this set, can be non-deductive. The above considerations suggest that there was *logica docens* 'at the beginning' and it became *logica utens* through application. According to this image, logic is thus applied in the same way as mathematical schemes in a concrete physical theory, e.g. Euclid's geometry in classical mechanics or non-Euclidean geometry in a specific theory of relativity. This circumstance makes a naturalistic interpretation of logic difficult, and even impossible, because, generally speaking, laws of logic are considered abstract to the highest degree and as such are thought to belong to Plato's world of forms.

A contemporary follower of Plato says that naturalism is helpless with respect to the domain of abstracts for two reasons. Firstly, because the naturalistic view acknowledges the existence of temporal-spatial objects as the only ones (there exclusively exist temporal-spatial and changeable objects), while the logical realm exists out of time and space. According to Platonism, this is the main ontological difficulty of naturalism. Secondly, the naturalist also faces an epistemological problem, since as a genetic empiricist with respect to sources of cognition he or she cannot elucidate the genesis of the genuine universal and infallible knowledge represented by logic and mathematics. In particular, the follower of Plato adds that no empirical procedure is able to generate logical theorems as true, independent of empirical circumstances. Platonism is – as a matter of fact – a historical and metaphorical label in above remarks. From the systematic point of view, it is much better to use transcendentalism (or anti-naturalism) as the opposition against naturalism, since every criticism of naturalism (sooner or later) makes references to transcendental arguments in the sense of Kant. It is in this way that, for example, criticism of psychologism (as a version of naturalism) was executed by Frege and Husserl; one could say the same about Moore's arguments against reducibility of axiological predicates to non-axiological ones. In general, transcendentalists reproach naturalists with what Moore defined as a naturalistic fallacy on the occasion of his criticism of reduction of moral values to utility. Dualisms of facts and values or logical and extra-logical theorems are not the only ones which naturalism has difficulties with. Other oppositions, from which – in the transcendentalists' opinion – naturalists are cut off in the sense of impossibility of their satisfactory explaining are, for example, the following: physical information – semantic information or quantity – quality.

This, of course, is fairly true that naturalism must meet various difficulties. The criticism of this view, however, overlooks problems of anti-naturalism, which Moore had already drawn attention to. The arguments he used were that the super-naturalistic (Moore used this qualification) grounding of morality as rooted in the supernatural world is a similar error to that of reducing axiological predications to ones definable in purely natural categories. Another problem of transcendentalism arises with respect to the so-called Benacerraf argument indicating the enigmatic character of cognition of mathematical objects, provided that each cognition consists in causal interaction on the part of the object of epistemic acts, while numbers – on the power of their nature according to Platonism – do not interact in a causative way with people. How, then, can an anti-naturalist explain the genesis of logic? He or she can either assume – as Plato did – that the world of abstract forms is eternal, or argue – like Descartes – that certain ideas are inborn, or still – like some theists – that man obtained logic as a gift from God when he was created as the *imago Dei*. The Platonic and Cartesian paths are ad hoc, whereas that of theists is based on extra-scientific premises. In any case, the situation of an anti-naturalist is not to be envied as it must resort to secret beings (souls, spirits, ideas) and secret kinds of cognition (intellectual intuition, etc.). The naturalist can paraphrase the title of Hoimar von Ditfurth's book *Der Geist fiel nicht vom Himmel* (*The Ghost Has Not Fallen From Heaven*) by saying that logic has not fallen from the other world, Platonic or others (as regards the defence of naturalism in other contexts compare Woleński 2016).

For a positive naturalist account of the genesis of logic, it is indispensable to combine the dichotomy *logica docens – logica utens* with the notion of logical competence, modelled on grammatical competence in Chomsky's sense. Both abilities play a similar role. The grammatical competence generates the right usage of linguistic devices, while the logical competence determines the application of logical rules in inferential processes. Nevertheless, the analogy is not complete, at least according to my own concept of the question. Whole Chomsky defines grammatical competence simply as grammar, the distinction which I am going to use differentiates logic, both theoretical and applied, from logical competence. The last category refers to a determined disposition of the biological organisms which are capable to perform mental functions (compare further comments below). Speaking more precisely, logical competence is the ability to use operation *Cn*. Thus, a logical theory is not logical competence, but its articulation. The dispositional character of logical competence does not settle whether each element of logic as a theory finds its coverage in its natural generator. By the way, a negative answer is rather obvious as the development of logical theories was and is heavily dependent on nature and needs of communicative interactions within human society. Further considerations in this paper will be devoted to the genesis of logical competence. They refer to the genesis of logic inasmuch as without the possibility to create and apply rules of logic there would not appear logic in either of the two distinguished senses. In other words, *logica docens* and *logica utens* are derivatives (precisely speaking – one derivative) of logical competence. This circumstance justifies the title "Naturalism and the genesis of logic". Anyway, one of the main theses of this paper says that logical competence is not eternal, it appeared in the Cosmos at some time and is rooted in the biological structure of organisms. However, I have to make it clear at once that I do not treat my comments as relating to the biological question as empirical. My cognitive interest is of the philosophical nature and remains within evolutionary epistemology. However, in compliance with my metaphilosophical convictions, I have to take into consideration the output of empirical sciences, biology, in particular, in the analysis of philosophical problems. Otherwise speaking, philosophical analysis, though somehow speculative in its character, is superstructured on empirical knowledge.

In accordance with the above explanations, logical competence precedes logic, both theoretical and applied; nevertheless, there is also a feedback because theoretical reflection on logic and its applications to concrete questions can enhance the logical competence. Everything points to the fact that the logical theory required a prior application of rules of logic and the development of language. In the case of the Mediterranean culture, the applied logic appeared, for sure, with Greek mathematicians and philosophers. When Anaximander said that there does not exist the principle of closeness, since it would create the boundary of *apeiron* which is boundless, he made use of a rule similar to *regressum ad absurdum*. Pythagoras proved the existence of irrational numbers and his reasoning was a proof by reduction in the modern sense. Various paradoxes formulated by the Eleats were of a similar character. The first logical theory, that is, Aristotle's syllogistics, originated much later, although on the basis of extensive practical material accumulated earlier. It was also, in a vital way, linked to the structure of sentences of the Greek language. One cannot, however, say that carrying out logical operations requires knowledge of the language because they are typical of infants (see Langer 1980), and the latter do not have linguistic material at their disposal yet.

The question of the sense of understanding in animals, other than humans, is controversial, yet the following example (which can be treated as an anecdote) is only too suitable in this place (see Aberdein 2008). In 1615, in Cambridge, there was a debate devoted to the dog logic, which was attended by King James I. The problem concerned the question whether hunting dogs which were used for locating game during hunting, applied logic, in particular the so-called law of disjunctive syllogism in the form "*A* or *B*, thus if non-*A*, then *B*" (this question had already been considered by Chrysippus). Let us suppose that a hound reaches a fork in the roads. The trailed game runs away to the right or to the left. The hound establishes that it is not to the left, and therefore runs to the right. The debate had a very serious character and a truly academic form. John Preston (a lecturer of Queens' College) defended the thesis that dogs apply logic, whereas his opponent, namely Matthew Wren (of Pembroke College), argued that hounds are directed solely by the scent and it is the only reason why they choose the right direction. The role of the moderator was played by Simon Reade (of Christ's College). When the latter acknowledged M. Wren to be right, the King, himself a great enthusiast of hunting and relying on his own hunter's experience, observed that the opponent, however, should have a better opinion about dogs and lower about himself. Wren, very skilfully managed to get out of the tight situation by saying that the King's hounds – in contrast to others – were exceptional, since they hunted upon the ruler's order. This compromising solution is said to have satisfied everybody. After all, even if hunting dogs do apply disjunctive syllogism occasionally, they certainly do not do this making use of a language.

The debate held in the presence of the King of England is a good illustration of a certain difficulty regarding an analysis of the genesis of logic. There appears the question of what evidence could help in this case. The debaters in Cambridge considered the dogs' behaviour and drew conclusions from that. In the case of humans, we can observe signs of inferential processes in people or base ourselves on the written evidence of the past. Anyway, the empirical base is greatly limited. Not much can be inferred from the inscriptions found on the walls of caves inhabited by our distant predecessors. All the information through which human logical competence manifested itself has been recorded in a language developed to a such degree that it made it possible to encode the deductions carried out factually, even if it did not suffice to formulate a logical theory.

In this respect, the genesis of logic appears to be more mysterious than the appearance of mathematics (see Dehaene 1997) or language (see Botha 2003, Johansson 2005, Larson, Déprez, Yamakido 2010, Talerman, Gibson 2012). In both mentioned domains, especially in the latter one, there have appeared a host of works and theories. In particular, the question of whether animals can count and make use of a language, or at least a protolanguage, has been discussed (see Hauser 1997, Bradbury, Vehrencamp 1998).

Studies in the origins of logic are limited to various inquiries into the logical competence of children going through their pre-language period or that of people living in primitive societies. This provides solely epigenetic and ontogenetic material, whereas phylogenetic is taken into account only to the extent in which the traditional and strongly speculative Haeckel's assumption that ontogenesis reproduces phylogeny is accepted. Nevertheless, considerations concerning the origins of the calculating and languafe competence are important also to the discussion on the origins of logic. This concerns, in particular, the concept of origin and the development of grammar and sign systems (see Heine, Kuteva 2007). According to a fairly common conviction, signs were the earliest to appear, especially expressive ones, then there were iconic signs, followed by symbols. This corresponded to the evolution of grammatical structures from nominal through sentential-extensional to sentential-intensional. Thus, the development of language progressed on the basis of s transition from a-semantic or little-semantic objects to fully-semantic ones (intensionality symbolism). The origin of language has always been the object of an animated interest on the part of philosophers (compare Stam 1976 for a review of earlier theories). In 1866, the French Linguistic Society decided that considerations on this subject should be excluded from sciences. As a matter of fact, a renaissance of studies of the appearance and evolution of language was observed in the mid-20th century. One can speculate that works dealing with the genesis of logic, if they had been written on a mass scale in the first half of the 19th century, would have shared the fate of linguistic dissertations on the origins of language, which were deemed as too speculative.

It is not without significance to model microbiological and neurological processes, for instance, through cell automata (see Ilachinski 2001), or even with the help of advanced mathematical techniques (see Bates, Maxwell 2005) and computational ones (see Lamm, Unger 2011). These enterprises indicate that the very organisms themselves, and whatever is happening inside them, possess properties which can be formulated mathematically. However, far-fetched methodological carefulness is indispensable. The title of one of the quoted books runs as follows: *DNA Topology*. It can be understood in a dual way: firstly, it suggests that, for instance, DNA has a looped structure in certain circumstances; secondly, this can be understood in a weaker manner, i.e. in such a way that a topological notion of a loop models certain properties of DNA. Reading the book by Bates and Maxwell inspires to conclude that the authors make use of both meanings. My opinion on this problem consists in recommending the other sense of modelling. It is only assumed here that the world is mathematizable (that is, describable mathematically) due to its certain properties, yet it is not mathematical. Works in the field of the evolution of language and those devoted to modelling biological phenomena, as a rule, accept naturalism, silently or explicitly. It is well-expressed by the appearance of biosemiotics (see Barbieri 2003, Barbieri 2008, Hoffmeyer 2008, Favareau 2009), cognitive biology (see Auletta 2011; this author declares the theistic *Weltanschauung*, yet suspends it in his book) or the more and more popular physicalization of biology (see Luisi 2006, Nelson 2008). Since syntheses of biology and semiotics or biology and cognitive science can be conducted, there is no reason not link logic to biology.

The only advanced attempt at the naturalistic grounding of logic that I am familiar with derives from William Cooper (see Cooper 2001), who considers the following sequence: (*) mathematics, deductive logic, inductive logic, theory of decision, history of life strategies, evolution theory. The relations between elements (*) are such that from the evolution theory to mathematics we deal with implication, whereas reduction proceeds in the opposite direction. As far as deductive logic is concerned, it is directly implied by inductive logic and reduces itself to the latter. The evolution theory is the ultimate basis, both for implication and reduction. Briefly speaking, deductive logic arose at a certain stage of evolution (Cooper does not make it precise in detail) through a natural selection and adaptive processes. Cooper's schema leaves a lot to be desired. Omitting the lack of a more detailed definition of the 'production' of logic through the process of evolution, which was indicated earlier, the notions of implication and reduction are not clear in Cooper's model. Since deductive logic (that is, $Cn\varnothing$) is implied by any set of sentences, the role of inductive logic (I neglect here disputes relating to its existence; however, see below) is not specific. In consequence, the reduction of deductive logic to inductive logic appears to be highly unclear. Moreover, the phrase 'logic as part of biology' (the subtitle of Cooper's monograph) is ambiguous. It may mean that *logica docent* is part of the biological theory (more precisely: the evolution theory) or, also, that deductive competence (Cooper does not use this name) is an element of the biological equipment of human being. Indeed, in compliance with the well-known maxim of Theodosius Dobzhansky, nothing makes sense in biology if it is not considered in the context of evolution, but this does not mean that everything can be explained on the basis of the evolution theory. Cooper, in his analysis, ignores genetics completely and this is, perhaps, the most serious deficiency of the model.

A purely evolutionistic classical approach towards the establishment and development of human mental competences, such as the ability to use a language or reasoning, is – in an outline – as follows (it can be found in countless publications dealing with the theory of evolution and its application to different specific problem areas; compare, for instance, Lieberman 2005, Tomasello 2010). The Universe appeared about 15 billion years ago (all the dates here are given in approximation). The age of our Earth is 4.5 billion years. The first cell appeared a billion years later, and multicellular organisms after the next 2.5 billion years. Plants have been around for 500 million years, reptiles – for 340 million years, birds – 150 million years, and apes – for 7 million years. The species of *homo* appeared two million years ago, *homo erectus* – from 1 million to 700 thousand, and *homo sapiens* – 200 thousand. The cultural-civilizational evolution marked with the language (in the understanding of our modern times), the alphabet and writing began 8.500 years ago. Three and a half billion years from the moment of the appearance of the first cell to that of the appearance of civilization and culture were completely sufficient to form the mind capable of performing typical intellectual activities, in particular, to carry out a logical operation. *Homo sapiens* must have been able to do so much earlier, maybe it happened at the beginning of this species. It cannot be ruled out that rudiments of logical competence had already been available to *homo erectus*. Establishing the date of the appearance of logical competence in the course of evolution, and attributing it to organisms other than human, does not seem particularly important here. A safe evolutionist hypothesis in this respect can claim, for instance, that the inferential ability appeared by way of a randomly acting mutation, and because it proved to be an effective adaptive tool, it was developed by *homo sapiens*, also by virtue to available and more and more perfect linguistic instruments. The logical theory appeared as the final product of the long evolution process. This is an adaptation of the classical concept of the evolution of language (compare, however, the conclusions at the end of the paper).

Neo-Darwinian evolutionism connects the appearance of life and its further evolution with entropic phenomena (see Brooks, Wiley 1986, Küppers 1990). This perspective leads to the need for indicating anti-entropic phenomena, i.e. mechanisms which maintain stability of organisms and their internal order, and thereby determine the continuance of their existence (see Kauffman 1993). The decisive event to enhance a serious revision of the evolution theory was the discovery of DNA structure by Crick and Watson in 1953 (the model of double helix), as well as further research into genetic encoding. Those results demonstrated the necessity of a more profound linking of evolution with genetics. The notion of genetic information and manners of its transferring became the key instruments of a new biological synthesis, obviously, while keeping suitably modified classical categories of the evolution theory. Notice that formal analogies between information and entropy caused biologists and philosophers of biology to become interested more closely in relations between the first notion and the course of biological processes since as early as in the 1920s, (see Yockey 2005).

Several facts established by molecular biology are significant from the point of view of this paper (for a while I mention them without a 'metalogical' commentary; I entirely omit the physical-chemical questions, likewise the mechanism of hereditariness). Firstly, passing genetic information is directed from DNA through RNA (more precisely: mRNA – the letter 'm' denotes that RNA is in this case a messenger, that is, an agent passing information) to proteins. This observation makes the so-called main dogma of molecular biology. There are, as a matter of fact, certain exceptions in this respect (e.g. in the case of viruses), but at least in the so-called eukaryotic organisms (humans belong to this biological group) transmission of information is in compliance with this dogma. Secondly, genetic information is passed in ordered, linear, discreet, and sequential a manner. Thirdly, DNA particles are subject to replication (copying) and recombination (regrouping). Fourthly, the intracellular information system encodes and processes information, which causes the encoding in question to be interpreted as a computational system and to be modelled accordingly. Fifthly, passing of genetic information is not deterministic but random in its very nature, which is why there may appear genetic novelties. This last fact is vital from the point of view of the evolution theory because it explains the way in which mutation appears on the microbiological level.

The view that genetic information is of the linguistic character is only too tempting. Indeed, it is very often that we can see it treated as a language. And thus, we can speak about alphabets, words, syntax, codes and encoding, or about translations (in the sense of transfer from the genetic language into another one); this is done especially by representatives of bio-semantics, who – in the genetic information – detect a semantic dimension or, at least, its germs. Such an approach is, however, very debatable (compare the discussion in Kay 2000; Sarkar 1996a rejects the notion of genetic code, but this solution seems too radical). In true fact, technical elaborations of genetics avoid comparing the genetic code with the language (see, for example, Klug, Cummings, Spencer 2006). Regardless of the applied language, for instance, some write about 'words' as components of the genetic code, surely using quotation marks to indicate a certain metaphorical investing of genetic information with the linguistic dimension, while others do so about words, we can easily find here the problem of relation of physical information as something quantitative to the semantic information as qualitative. The mathematical theory of information concerns the former, and only indirectly relates to the latter. The well-known Shannon's statement on the capacity of channels of transmitting information and limiting the so-called information noise has a sense only with reference to its quantitative understanding. The genetic information is a kind of physical, not semantic information. On the other hand, processing the former, i.e. quantitative, into the other, i.e. qualitative, is a notorious fact, for example, reading a book – as long as we understand the language in which it has been written, we rapidly process the physical stimulus into semantic units, i.e. such that we understand according to their linguistic sense. For the time being, we do not know the mechanism of this transformation and it makes the biggest anthropological puzzle (see Hurford 2007). Perhaps, the properties of the genetic information lie at the foundations of what may be called semiotization of mental processes, yet this is a fairly speculative assumption from the biological assumptions, though one could consider it as philosophically justified to some degree.

At first sight, if the genetic information were a language in the full or merely approximated sense or, we could look for the genesis of logical competence directly on the microbiological level. After all, the properties of the genetic code, whatever it is, stand far from those that can serve to define operation *Cn*. Nevertheless, these properties can be tied to logic in their understanding of today. Before I pass on to show this relation, I will draw attention to certain theoretical questions. Kazimierz Ajdukiewicz (see Ajdukiewicz 1955) divided inferences into deductive, increasing the probability (inductive in a broad sense) and logically worthless. The first are based on operation *Cn* which holds between the premises and the conclusion (the conclusion results logically from the accepted assumptions), the second increase the probability of the conclusion on the basis of the premises, and the third ones are devoid of any logical relation between the links, e.g. 'if Kraków lies on the Vistula, then Paris is situated in France'. Logic, in this context, is understood in a broader way than at the beginning of the paper, since it includes also induction rules. We can, too, extend respectively the notion of logical competence, but I do not wish to consider such a generalization. Treating the thing from the information point of view (see above), while deduction does not broaden the information included in the premises, although it does not allow it to be lost, the conclusion is false, which disperses (in the sense of entropy) the information acquired earlier, whereas logical inference that is worthless is redundant from the information point of view.

The infallibility of the rules generated by operation *Cn* derives from the fact that they correspond to theorems of logic, i.e. to sentences (formulas) that are true in all circumstances. One of the axioms of probability calculus is the assumption that there exists an event whose probability obtains the value 1 (the whole space on which the probability measure is defined constitutes this event). An interesting interpretation of this axiom consists in acknowledging that it prevents levelling (dispersing) of probabilities ascribed to particular occurrences, that is subsets of the whole space. In other words, this axiom saves the differences in the amount of information, which conditions its flow. Thus, it performs the anti-entropic function, i.e. blocks the dispersion of information, it protects it in this way. Operation *Cn* can be understood also as an instrument of protection of information from its dispersion, since it prevents the formation of false information on the basis of true information. As I have already mentioned, logic of induction is debatable, yet – on the other hand – nobody contradicts the fact that at least certain induction rules, e.g. those of statistical induction, are rational. It is true that they do not exclude the dispersion of information, but they are still able to somehow normalize its flow and in this way save or control it. Inferences that are logically worthless do not play any role in the processes of information protection.

Saving information (obviously it is not problems of the legal or moral nature that I mean here), both physical and semantic, appears as a vital function of all organisms, which operate with a given type of code. Since we treat operation *Cn* as an information-protective instrument, saving the possessed content (in the sense of information content, not necessarily meaningful in the sense of intensional semantics), then – at least from the naturalistic point of view – the logical consequence has the biological rooting. In relation to this issue, I will return to the properties of the genetic codes and genetic information mentioned earlier, this time in the metalogical context. Here are the features of the genetic code (see Klug, Cummings, Spencer 2006, p. 307; I keep abstracting from the nature of elements of the code with one exception of amino acids due to the comprehensiveness of certain formulations): (i) genetic code is written in the linear form; (ii) if we assume that mRNA consists of 'words', then each of such words has three 'letters'; (iii) each three-letter group, that is, the codon, determines another element in the form of an amino acid; (iv) if the code is unambiguous, it delineates one and only one amino acid; (v) if the code is degenerated, the given amino acid can be determined by more codons; (vi) the genetic code includes the initial signal and the terminal one in the form of codons initiating and finalizing the processes of passing the genetic information; (vii) the genetic code does not contain punctuation characters (commas); (viii) elements of the genetic code do not overlap, that is, a concrete 'letter' can be part of only one codon; (ix) the genetic code is nearly universal, i.e. apart from few exceptions, the same 'dictionary' of encoding serves all viruses, procariotic and eucariotic organisms. Completing the remarks offered earlier, I would like to add (see Klug, Cummings, Spencer 2006, pp. 264 –265) that replication of DNA (which forms two new strings of the primary helix) can be semi-conservative (each replicated particle of DNA has one old string and one new one), conservative (the parent string is conserved as a result of synthesis in two new strings), or dispersed (old strings are dispersed in new ones). The most frequent is the case of semi-conservation. Nevertheless, the genetic information that exists earlier is inherited by helixes formed by way of replication.

It follows from properties (i) – (ix) that the 'syntax' of the genetic code is rigorous. It is based on a detailed specification of simple elements ('letters') and their combinations (codons). The lack of commas points to the fact that it is a series of concatenations. A code is unambiguous inasmuch as it is not degenerated. This property can be likened to syntactic correctness, while degeneration to a lack of it. 'Letters' are atoms in the same sense as a simple expression, which is non-decomposable any further. Transformation of a codon into an amino acid is a function, unless the code is degenerate. The beginning and the end of the procedure realized by the code is clearly marked with separate 'words'. I marked some expressions with letters so as not to suggest treating the code as a language. The linguistics-oriented terminology could easily be avoided through speaking about configurations and their elements. Genetic codes treated in this manner can be, and are, similar to electric nets or cellular automata, which – as a matter of fact – is underlined by above-mentioned modelling of genetic phenomena. The essence of things relies on that the outline 'syntax' is of an effective character and is trivially recursive, since operations realized by the codes are of the terminal character.

Although there are similarities between the structure of the genetic code and the syntax of formal languages, there is no reason why we should see the genetic concatenation as the result of the operation *Cn*. On the other hand, if we were to consider the configuration determined by codons, it is clopen in the topological sense, which is also characteristic of the space of sentences on which the logical consequence operates together with non-deductive rules of semantic information organization. The semi-conservative character of the most typical replication of DNA corresponds to this. Closing of a part of this space protects the information accumulated earlier, and the fact that it includes also open subsets secures the appearance of new information. Sometimes it is said (see Kauffman 1993, pp. 203, 447–449) about the channeling of processes of genetic regulation through extensional (Boolean) functions. Let x be an active element in such a process, and object (x or y) a regulated element. Then, the object (x or y) is also active. The procedure, in this case, is analogous to that applied in the synthesis of electric networks. In the terminology used in this paper, the channeling (in the sense of Kauffman) is a partial objective case of logical consequence. A general conclusion which can be derived from the registered analogies is as follows: the genetic code is the biological foundation of logical competence. Since speculation becomes a philosopher, the thing can be framed as follows: topological or proto-topological properties of the 'genetic space' directed the biological evolution in such a direction that it developed – perhaps by way of relevant mutations, with the appearance of dispositions to operate with logical consequence.

It is not a feasible thing to establish various vital details. It is not known when logical competence appeared in its fullest beauty, so to speak, and what its scope is with reference to other species than ours. Putting it differently, it is not known whether the logical ability is only granted to humans, or also to other species, and even to what extent the real human logical competence corresponds to its abstract image formulated in the logical theory. Perhaps, the evolution theory could add something relevant in this matter. It seems, for instance, that species which invest in a lower number of offsprings (birds and mammals) have been 'forced' to create stronger means of protection of genetic information than insects, reptiles and fish, in which dispersion is compensated with a great number of potential specimens. Be it as it were, a naturalist claims, to repeat it once again, that logical competence has not fallen from the other world but came into being on the planet Earth. This makes, similarly as other mental operations, a realization of dispositions determined by the genetic equipment and the course of the evolution. If it is inborn, then it is so phylogenically, not ontogenically.

Acknowledgements. I would like to thank Professor Janusz Bujnicki and Professor Adam Łomnicki for their comments that toned down the temper of a philosopher who interprets results of biology, and Professors Urszula Wybraniec–Skardowska, Janusz Czelakowski (in particular) and Roman Murawski for their comments and explanations pertaining to the questions of analogies between logic and topology.

VII
Some Analogies between Normative and Epistemic Discourse

Kant divided sentences (I will use the terms 'sentence' and 'propositions' as referring to the same entities) into assertoric, problematic and apoidectic. This division was traditionally considered as very closely related to modals, that is, linguistic words expressing modalities. In particular, problematic propositions express possibility, but apoidectic – necessity. Even if we consider assertoric force as modal (it means that assertoric propositions are regarded as modal), Kant's account was entirely restricted to so-called alethic modalities and, thereby, it was very narrow. According to the contemporary view, modal sentences, not modal words, are the most basic unit of the modal discourse. The variety of modal sentences is much wider than necessities (utterances expressing necessity) and possibilities (utterances expressing possibility) and includes their various kinds (or species), except paradigmatic alethic modalities (see Portner 2009 for a survey from the linguistic point of view): for instance,[4] deontic (it is obligatory, permitted, forbidden, etc.; I will also use the adjective 'normative' as referring to equivalents of normative modalities), axiological–ethical (it is good, wrong, ethically neutral, etc.), axiological-aesthetic (it is beautiful, ugly, aesthetically neutral, etc.), epistemic (I know, I do not know, it is not the case that I do not know, etc.) or doxastic (I suppose, I do not suppose, it is not the case that I do not suppose, etc.; the same goes for other personal denominations). We have a lot of another doxastic modalities, for example, I guess, I think, I assert, I consider, I believe, I think, I maintain, I assess, I doubt, I expect, etc.

A natural question is: are there similarities between particular species of modalities? Although my main focus concerns deontic and epistemic modalities, comparisons with other kinds of modals will be taken into account as well. The main part of the paper investigates formal issues, but I shall discuss some substantial problems at the end of this paper. Some further preliminary explanations are in order here. I will not distinguish norms and normative propositions. All sentences considered in this paper are understood as indicative in the grammatical sense. This entails that they are true or false. Deontic modalities form a

4 The list is incomplete, as it neglects, for example, erotetic (related to acts of questioning) and temporal modalities (some additional illustrations will be mentioned later).

relatively homogenous class, even if we distinguish categorical and hypothetical deontic modals. The situation is different in the case of epistemic and doxastic modalities. The division between these two species is related to the ancient distinction between *episteme* (knowledge) and *doxa* (opinion). However, a formal analysis of epistemic and doxastic modals does not depend upon differences, if any, between *episteme* and *doxa.* As I will later argue, one logical feature of the modalities in question suggests grouping them into one general category. Thus, I will simply speak about epistemic modalities in the broad sense as covering 'I know' (the epistemic or epistemological modality in the restricted sense) as well as the traditional doxastic modals. Epistemic modalities analysed in this paper express so-called propositional attitudes having a cognitive role. Thus, I disregard other sources of attitudes, for instance, religious or magical. I also neglect perceptual modalities (note that this qualification is used by psychologists), like I see, I hear, I touch, etc. They are epistemic in the broad sense, but provide some special problems, which exceed the scope of this paper.

I will use a very elementary logical machinery, namely the logical square and its two generalizations. The present analysis follows my earlier papers (see Woleński 2008b). It was Robert Blanché (see Blanché 1953) who proposed modal extensions of the logical square (see also McNamara 2007 for the deontic case; see Gabbay, Woods 2006, Blackburn, van Benthem, Wolter 2007 and Horsten, Pettigrew 2011 for advanced surveys of contemporary modal logic). The most popular version of the logical square (or the square of oppositions) records various logical relations between so-called categorical sentences. These sentences are as follows: every *S* is *P* (*SaP*), no *S* is *P* (*SeP*), some *S* are *P* (*SiP*), some *S* are not *P* (*SoP*). These relations are visualized by the diagram (**D1**):

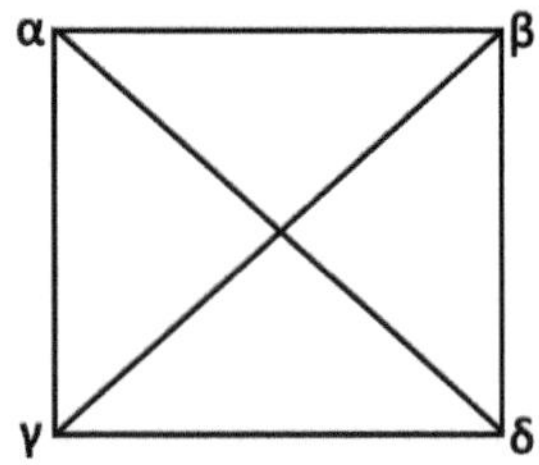

The point α can be read as $\forall xPx$ (for every *x*, *x* is *P*), the point β as $\neg\exists xPx$ (or $\forall x\neg Px$; there is no such *x*, that *x* is *P*), the point γ as $\neg\forall xPx$ (or $\exists x\neg Px$; there is such *x*, that *x* is not *P*). Let us understand α as $\blacksquare A$, β as $\blacksquare\neg A$, γ as $\exists xPx$ (there is such *x* that *x* is *P*), and the point δ as $\blacklozenge A$ and δ as $\blacklozenge\neg A$, we have the following logical connections (the symbol $\vdash A$ means 'it is provable in the logic generated by **(D1)** that *A*' or '*A* is a theorem in the logic generated by **(D1)**'; *A* is a formula of the propositional calculus:

(1) $\vdash \neg(\alpha \wedge \beta)$	(α and β are contrary);
(2) $\vdash (\alpha \Rightarrow \gamma)$	(α entails γ; γ is subordinated to α);
(3) $\vdash (\beta \Rightarrow \delta)$	(β entails δ; δ is subordinated to β);
(4) $\vdash (\alpha \Leftrightarrow \neg\delta)$	(α and δ are contradictory);
(5) $\vdash (\beta \Leftrightarrow \neg\gamma)$	(β and γ are contradictory);
(6) $\vdash (\gamma \vee \delta)$	(γ and δ are complementary);
(7) $\vdash (\blacksquare A \Leftrightarrow \neg\blacklozenge\neg A)$	($\blacksquare$ is definable as $\neg\blacklozenge\neg A$);
(8) $\vdash (\blacklozenge A \Leftrightarrow \neg\blacksquare\neg A)$	($\blacklozenge$ is definable as $\neg\blacksquare\neg$)
(9) $\vdash (\beta \Leftrightarrow \blacksquare\neg A)$	(β is definable as $\blacksquare\neg A$;

Due to (7) – (9), negation standing before $\blacksquare$ or $\blacklozenge$ is always eliminable. The above treatment of the box ($\blacksquare$) and the diamond ($\blacklozenge$) is purely syntactic and does not provide too many semantic and pragmatic intuitions. However, it is not difficult to see that (1) – (9) agree with the colloquial meaning of modal words. Alethic modalities are perhaps the most instructive example, but it is easy to check that also deontic and epistemic modalities behave according to the principles generated by **(D1).** We can also note a linguistic fact, namely that deontic and alethic interpretations have their own linguistic labels at γ (possibility, permission), but other do not. In particular, we have no separate word for the epistemic possibility (I do not know that $\neg A$).

The diagram **(D1)** can be extended to logical hexagon **(D2)**

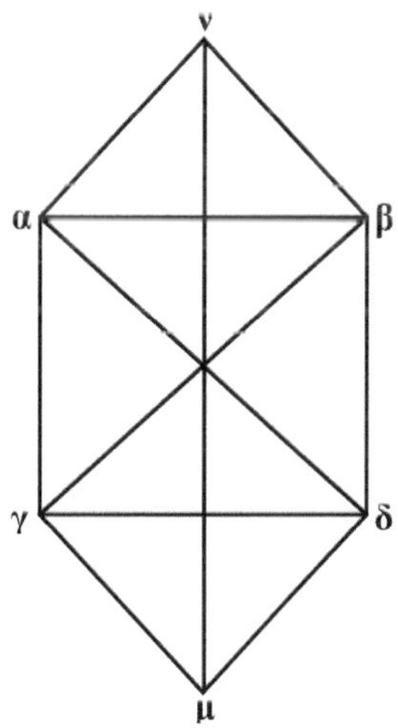

The point μ defined as $\blacklozenge A \wedge \blacklozenge \neg A$. This formula can be read alethically as 'it is possible that *A* and it is possible that not-*A*' (*A* is accidental), deontically as 'it is permitted that *A* and it is permitted that not-*A*' (*A* is indifferent) and epistemically 'I do not know that *A* and do not know that not-*A*'. Further, we have $\nu =^{df} \blacksquare A \vee \blacksquare\neg A$ (I postpone comments about the meaning of this formula). We have (among other things):

(12) $\neg(\nu \Leftrightarrow \mu)$ (ν and μ are contradictory);
(13) $\alpha \vee \beta \vee \mu$ (α, β, μ are mutually exclusive and jointly exhaustive).

Thus, the logic generated by **(D2)** is richer than the one obtained from **(D1)**. Historically speaking, it was Robert Blanché (see Blanché 1953) who proposed modal extensions of the logical square (see also McNamara 2007 for the deontic case).

The extension of the square **(D1)** to **(D2)** increases the expressive power of our modal discourse and allows to express some important facts. For instance, we can distinguish two senses of contingency, one understood as non-necessity (the point δ) and another as expressed by μ. The latter is sometimes conceived as two-sided possibility as contrasted with one-sided possibility (the point γ). Exactly analogical principles hold for deontic and epistemic modalities. In particular, we can distinguish permission and indifference in the case of deontic modals. Clearly, due to (2), obligation entails permission (if *A* is obligatory, it is also permitted) but this implication fails if permission is understood as indifference. In fact, most people consider permission rather as expressed by indifference than by the point (γ). The formula (13) asserts, freely speaking, alethically that every fact is either necessary or impossible or accidental, but deontically, that every action is either obligatory or prohibited or indifferent. Similar facts can be stated for epistemic modals, although their formulation is less intuitive for the absence of special single words expressing the epistemic counterpart of alethic possibility. Note that logic as such does not decide whether something is necessary, possible, accidental, obligatory, permitted, known, etc. Hence, for instance, deriving the existence of necessary beings or necessary duties from the fact that there are contingent (or accidental) entities or permitted (or indifferent) actions is implausible on purely logical grounds and requires additional, extra-logical premises. This fact is neglected by many proponents of demonstrations of God as a necessary being. In fact, ontology admitting contingent beings as the only furniture of the world (the realm of existence) is perfectly consistent.

Note that I employed the predicative reading of modals in some foregoing explanations, for example, '*A* is necessary' instead 'it is necessary that *A*' or '*A* is obliged' instead 'it is obligatory that *A*'. Using the traditional terminology, I switched from the *de dicto* interpretation of modal sentences (modals function as propositional or sentential operators and occur before their arguments) to the *de re* reading (modal operators are replaced by corresponding predicates standing inside sentences). There are various important problems here. I mention two such questions. Firstly, the status of the letter A must be re-interpreted under the *de re* interpretation, because we cannot consider it as a sentential variable. For instance, the phrase 'tomorrow will rain is possible' is grammatically incorrect and should be replaced by something like 'tomorrow rain is possible' or 'that tomorrow will be rain is possible'. Freely speaking, the predicative reading of modals refers to the nominal denomination of facts, situations or states of affairs. Secondly, it is controversial whether modal predicates refer to properties of objects, facts, etc. Let us only mention that traditional ontology considered alethic modalities as expressing modes of beings and not as words contributing to the content of non-modal phrases. Thus, if I say 'A is necessary', the adjective 'necessary' does not increase the semantic content of A; the situation is, of course, similar, in the case of the sentence 'it is necessary that A'.

We can additionally increase the expressive power of (**D1**) by adding quantifiers. It is particularly suitable for epistemic contexts. Consider the formula $\forall A(\blacklozenge A \wedge \blacklozenge \neg A)$ under the epistemic interpretation with 'I know'. It means that for any A, *I* do not know that $\neg A$ and I do not know A. This statement is tantamount to saying that I am completely ignorant, that is, I have no knowledge about any fact. On the contrary, the formula 'for any A, a knows that A or a knows that $\neg A$' qualifies that a is omniscient. Correspondingly, the formulas $\exists A$(a knows that A) and $\exists A$(a does not know that $\neg A$ and a does not know that $\neg A$) express, respectively, that a knows something, and a is ignorant with respect to something. Now, read the box as 'A is determined'. Thus, the formula $\forall A(\blacksquare A \vee \blacksquare \neg A)$ means that everything is determined and perhaps can be considered as an adequate statement of radical determinism. Its denial, that is, the formula $\forall A(\blacklozenge A \wedge \blacklozenge \neg A)$ expresses a radical version of indeterminism (everything is accidental). Clearly, the conjunction $\exists A(\blacksquare A \vee \blacksquare \neg A) \wedge \exists A(\blacklozenge A \wedge \blacklozenge \neg A)$ can be accepted by moderate determinists as well as moderate indeterminism. The deontic reading of the universal closure of v represents a dream of every bureaucrat that everything is obligatory or prohibited. On the other hand, the deontic reading of the formula $\forall A(\blacklozenge A \wedge \blacklozenge \neg A)$ expresses the situation of a completely anarchistic world in which everything is indifferent (there are no obligations or prohibitions). We can also consider an axiological (ethical, but also aesthetic) interpretation. The point α means that a is good. If we apply the universal quantification, we obtain the statement that any object is good. The medieval theory of transcendentals accepted the axiom *ens and goodness convertuntur* (being and goodness are mutually interreducible, that is, extensionally equivalent). However, it is an extralogical statement, perhaps true under some additional philosophical premises, but certainly not evident in itself. A similar remark concerns moral pessimism which claims that everything is wrong and moral indifferentism (nothing is good or wrong) as well as the universal closure of ν providing a formal principle of moral manicheism (everything is good or wrong; Goodness or Evil). On the baseis of logic, we can only say that everything is good, wrong or morally indifferent or something similar, for instance, that everything is good or not-good. This simple fact should be recommended for everybody who discussed the question whether axiological or normative predicates refer to objective properties of objects or subjective accounts of them. Finally, let me observe that the box can be also read metamathematically as 'it is provable that *A*'. Assume that "provable" means "provable is a system *S*". Then, the formula $\forall A(\blacksquare A \vee \blacksquare\neg A$ means '*S* is deductively complete', but the formula $\exists A(\blacklozenge A \wedge \blacklozenge \neg A)$ – there are unprovable sentences in *S*.

(**D2**) can be further extended to the octagon (**D3**):

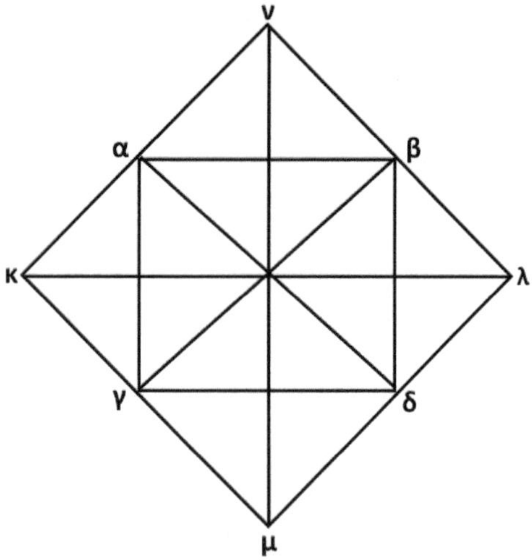

We put *A* and $\neg A$ at κ and λ respectively. Using Tarski's biconditionals, we have $A \Leftrightarrow \mathrm{T}A$ and $A \Leftrightarrow \mathrm{T}\neg A$, where T stands for 'it is true'. This addition allows to treat *A* and $\neg A$ as hidden alethic modalities. Observe that these new formulas are not available in (**D2**), unless we define *A* as $\blacksquare A$ or $\blacklozenge A$, but this would mean that box and diamond collapse to assertoric modals. Moreover, if F stands for 'it is false', we cannot prove that FA is equivalent to T$\neg A$. If we assume that the entire construction presented by **(D2)** is formally governed by classical logic, then the pairs $\{\blacksquare A, \neg A\}$and $\{A, \neg A\}$ are internally inconsistent. However, the situation changes if we take, for example, the Łukasiewicz or Kleene three-valued logic as the basic formalism, because if the variable A s valued by ½ (the neutral value), its negation possesses the value. Consequently, A and $\neg A$ are not inconsistent. In order to fix the standpoint in this paper, I assume that classical logic is our framework. Thus, as far as the issue concerns truth and falsity, we have the principle of bivalence (every sentence is either true or false) and, consequently, the equivalences F$A \Leftrightarrow$ T$\neg A$ and $\Leftrightarrow \neg$TA. The facts about truth and falsity are nor particularly important in my further considerations.

The importance of **(D3)** consists in the fact that it allows to account logical differences between alethic modalities on the one hand, and deontic and epistemic modalities on the other. In particular, the formulas, (#) $\blacksquare A \Rightarrow A$ and (##) $A \Rightarrow \blacklozenge A$ are valid in normal (that is, alethic) modal logic, but they fail in non-normal systems, like deontic and epistemic logic. In fact, if we read A as 'A obtains', we cannot accept that A obtains, when A is obligatory, because people violate their duties. The same holds for the epistemic interpretation (with some problems discussed below). On the other hand, clearly, if A is necessary, A obtains. Similarly, there is no entailment from deontic or epistemic possibility of A to obtaining that A; the reverse dependencies also do not hold. In order to capture these facts (plus concerning the points ν and μ) in a symbolic manner, we add:

(14) $\neg \vdash (\blacksquare A \Rightarrow A)$;
(15) $\neg \vdash (\blacklozenge A \Rightarrow A)$;
(16) $\neg \vdash (A \Rightarrow \blacksquare A)$;
(17) $\neg \vdash (A \Rightarrow \blacklozenge A)$;
(18) $\neg \vdash (A \Rightarrow \nu)$;
(19) $\neg \vdash (\nu \Rightarrow A)$;
(20) $\neg \vdash (A \Rightarrow \mu)$;
(21) $\neg \vdash (\mu \Rightarrow A)$.

Since we have $\vdash \blacklozenge t$ (t stands for a tautology) in deontic and $\blacksquare A$ in epistemic logic (see comments below), we should assume that the formula A is not tautological in (15), (16) and (17). A similar restriction concerns (14), (19) and (21), because tautologies follow from arbitrary premises. I will collectively call (14)–(21) generalized Hume's principles or theses (see Woleński 2006).

Why generalized Hume's principles? Consider the famous passage from Hume's *Treatise on Human Nature* (Hume 1951, p. 469):

> I cannot forbear adding to those reasonings an observation, which may, perhaps, be found of some importance. In every system of morality, which I have hitherto met with, I have always remark'd that the author proceeds for some time in the ordinary way of reasoning, and establishes the being of a God, or makes observations concerning human affairs; when of a sudden I am surpris'd to find, that instead of usual copulations of propositions, *is*, and *not*, I meet with no proposition that is not connected with an *ought*, or *ought not*. This change is imperceptible; but is, however, of the last consequence. For as this *ought*, or *ought not*, expresses some new relation or affirmation, 'tis necessary that it shou'd be observ'd and explain'd; and at the same time that a reason should be given, for what seems altogether inconceivable, how this new relation can be a deduction from others, which are entirely different from it.

The most important part of this quotation is a counterpart of (14) and denies that ought-sentences are derivable with is-sentences. More accurately, we should say that the former are not entailed by statements without words indicating deontic modals. This principle can be immediately reversed and generalized to other deontic modals. The next generalization extends Hume's principles to epistemic operators. That sentences with traditional epistemic and doxastic modals obey (14) – (21) is an important reason to group them into one general category.

There is, however, a problem with (14) as far as the matter concerns its epistemic reading. Consider the classical definition of knowledge, according to which knowledge is true justified belief. More specifically, we adopt the following equivalence:

> (*) I know that *A* if and only if (a) I believe that *A*, and (b) *A* is true, and (c) *A* is properly (sufficiently) justified.

Clearly, (*)(b) justifies that ■*A* entails *A*, contrary to (14). The first observation is that logical compulsion suggested by **(D3)** prevails over (*) as a definition governed by our intuitions. However, there are more powerful arguments against (*), even if we disregard so-called Gettier's counterexamples. Consider any past scientific theory, for instance, that of Copernicus. He calculated that planetary orbits are circular. Kepler demonstrated that planets move on ellipses. Thus, Copernicus believed in something false. Consequently, we should conclude that his heliocentric theory did not express knowledge. On the other hand, this conclusion seems absurd. Moreover, because most past scientific theories required corrections sooner or later, we should deny that even the greatest scientists achieved no cognitive successes. We encounter here a serious clash between the fact that our cognitive activities are mistaken and full of errors, and our initial intuition that knowledge as such excludes cognitive mistakes.

I think that the entire problem stems from the traditional concept of *episteme* as the ultimate knowledge. Plato, contrary to many opinions, did not accept (*)

unconditionally. He proposed that (c) should be strengthen by adding that we properly use our cognitive capacities, that is, necessarily intuit forms by reason. He also observed that *doxa* can be better or worse justified. Yet Plato denied that even a "good" *doxa* provides pieces of knowledge. The ancient Skeptics and Academicians took another stance, when they argued that since *episteme* cannot be achieved, we must remain with *doxa.* Otherwise speaking, they rejected (*)(b) and argued that truth is not accessible. Many later proponents of similar views suggested that we should weaken (*)(b) by saying that A is partially or relatively true. Yet there is some way to reconcile both views. In order to show that I will use a metamathematical example. What is a relation between truth and provability? The Löb theorem gives an answer for arithmetic of natural numbers (**AR**). It says that **AR** $\vdash (\vdash A \Rightarrow A$. Informally speaking, if something is provable in arithmetic, it is true. Now, assume, using the Gödel incompleteness theorem, that A is an undecidable arithmetical statement, for example, 'Arithmetic is consistent'. If we add A as a new axiom to **AR**, we obtain a consistent theory, let say **AR'**. Another consistent theory, let say **AR"**, can be build by adding $\neg A$ to **AR**. This new theory has the sentence 'Arithmetic is inconsistent' as an axiom. **AR"** is consistent, although ω-inconsistent. By the Gödel-Malcev completeness theorem (every consistent theory has a model), **AR"** has a model in which the assertion about inconsistency of **AR** is true. Since every axiom is logically provable, the same concerns the sentence in question. Thus, $\neg A$ is provable and true. Thereby, it is not a counterexample for the Löb theorem. Yet A and $\neg A$ cannot be true in the same model. In fact, A is true in the standard model, but its denial is true in non-standard models.

The above consideration about truth and provability in arithmetic may be paraphrased to knowledge in general. If we look at the history of science (I do not discuss the humanities), every époque generated something which can be regarded as the standard model or picture of the reality. It was formed by prevailing meanings of used words and related to justification procedures. If **Th** was a theory accepted in a given moment, it suggested its standard model. Other models were considered as non-standard or could be so qualified by later historians of science. For instance, if someone would say to Copernicus that planetary orbits were elliptical, his answer, phrased in terms of this paper, could be "well, but it is a non-standard view, although possible". On the other hand, a post-Keplerian historian of astronomy has reasons to say that Copernicus' picture was non-standard as compared with Kepler's theory, although consistent with the data available for Copernicus. As a result we have the following statement: if **Th** justifies A, then A. Yet we should not say that A is true *simpliciter*, but that it is true in a standard model of **Th**. Now, we can replace (*) by (it is better to use an impersonal formulation):

> (**) It is known that A if and only if (a) A is believed, and (b) A is true in a standard model generated by the given state of knowledge (science) and (c) A is properly (sufficiently) justified.

One can ask how standard models are related to the real world. The answer strongly depends on a view about scientific theories. In particular, the instrumentalist will say that this question appears as illegitimate, because theories or conceptual systems are our constructs, which are neither true nor false. The realist, in turn, maintains that standard models actually approximate our real world.

I will now return to formal-logical matters. Although (**D1**) – (**D3**) help us clarify several concepts, the expressive power of the related logical machinery is limited. The reason is that ■ and ♦ are unary operators; additionally, we exclude iterations of modals, for example, formulas like ■■A or ■♦A. Thus, we can neither formulate such principles as, for example, ■($A \Rightarrow B$) ⇒ (■A ⇒ ■B) (the law of monotonicity for ■) nor express characteristic axioms for the systems **S4** (■A ⇒ ■■A) or **S5** (♦A ⇒ ■♦A). Consequently, we should say that all considered diagrams display some unary modal logics without iterations of modalities. The diagrams (**D1**) and (**D2**) generate the so-called **D**-system, because the formula ■A ⇒ ♦A is its characteristic axiom, but the diagram (**D3**) is associated with **M** (or **T**), as it validates (■$A \Rightarrow A$). The formula ■t sometimes serves as a deontic axiom. It states (metalogically speaking) that something is obligatory. However, we should also admit the anarchistic world, that is, the world without obligations in which everything is permitted. The formula ♦t is more proper as a general axiom. It asserts that tautologies are permitted, that is, not prohibited. Assume that a tautology, for instance, $A \vee \neg A$, is not permitted. By (1), we have ■($A \wedge \neg A$), which is a contradiction. The formula ■t holds for cases in which something is obligatory. To see this, suppose ¬■($A \vee \neg A$). By (4) and the definition of ♦ by ■, we obtain ♦($A \wedge \neg A$), which is true in the case of the anarchistic world, because if nothing is obligatory or prohibited, everything is permitted. However, since the anarchistic deontic appears as unrealistic, one can assume that there are duties created by normative orders. This assumption generates a class of possible worlds to which domains belong that can be treated as standard relatively to a system of norms or knowledge. If **N** is a given normative order, ■A is true in our world if and only if A is true in all deontic alternatives (standard deontic worlds) with respect to the real world, that is, worlds in which all obligations are satisfied.

What about the epistemically anarchistic world, that is, world of which we (not subjects) have knowledge? Otherwise speaking, we consider the state of complete ignorance (nothing is known). Similarly, we should treat such situation as utopian and, thereby, non-standard. Consequently, (14) holds for deontic and epistemic sentences in a restricted form: for any standard world, $\blacksquare A \Rightarrow A$. Now, consider whether we should accept $\blacklozenge t$ or $\blacksquare t$ as an axiom of epistemic logic. Assume the first alternative. It entails that the formula $\blacklozenge(A \wedge \neg A)$ (epistemically interpreted) holds in the anarchistic world, because if it is not a case that for any A, A is known or $\neg A$ is known, then everything is epistemically indifferent. However, if we seriously take (**) or even the traditional definition of knowledge, we must reject the formula $\blacklozenge(A \wedge \neg A)$, because it means that we can know contradiction, that is, that they are true in some worlds, more precisely, in the anarchistic world. Thus, we should adopt $\blacksquare t$ in epistemic logic, similarly as in the case of **V** and necessity. In general, assumptions that logical truth and necessity are merely possible is too weak, because it does not exclude that inconsistencies and impossibilities might happen. Incidentally, considerations about the epistemic situation throw some light on the problem of the epistemic status of logic. Adopting $\blacksquare t$ means that we exclude the situation in which countertautologies are known, that is, believed, true and subjected to a justification (similar remarks concern other epistemic states). Since the laws of logic are universally valid, that is, they hold in all domains, we do not need to appeal to standard models.

We can note other dissimilarities inside the variety of modals, which satisfy Hume's principles. Deontic sentences are closed under replacement by logical equivalents. Otherwise speaking, if $\blacksquare A$ and B is logically equivalent to A, then $\blacksquare A$. This does not locate the deontic $\blacksquare$ in the same group as the necessity operator. The epistemic $\blacksquare A$ does not satisfy this property. For example, if I know that A and A is logically equivalent to B, it can actually happen that I do not know that B. Thus, epistemic contexts are strongly intensional, but deontic and alethic sentences are weakly intensional. The context $\mathbf{V}A$ is extensional, because if $\mathbf{V}A$ and B is true, then $\mathrm{T}B$ is true as well, and if $\mathrm{F}A$ and B is false, then $\mathrm{F}B$ (it was Carnap who distinguished these three cases; see Carnap 1947, pp. 46–68). Accordingly, there is no full correlation between extensionality and the principle (#). Although (#) is satisfied by T and the necessity box, the former is extensional, but the latter lacks this property. On the other hand, the deontic box and the necessity box are both weakly intensional, but (#) is invalid for the former. Finally, although epistemic sentences and deontic sentences satisfy Hume's principles, the former are strongly intensional, but the latter are not (they are weakly intensional). This situation has affinities with the closure problem (see Barke 2002). The box interpreted as truth, necessity and obligations forms a context which is closed under the consequence operation. Freely speaking, logical consequences of truths are true, logical consequences of necessities are necessary and logical consequences of obligations are obligatory. In order to make an epistemic context closed by the consequence operations, we have to assume that cognitive subjects know everything. In order to block the conclusion that *a* might not know that *B*, although he or she knows that *A* and *A* is logically equivalent with *B*, one must assume that omniscience concerns logic.

Fortunately, the mentioned dissimilarities do not prevent substantial comparisons of various modalities, as formal analogies are not enough for saying that two items, say **I** and **I'**, are somehow substantially (that is, by their content) related. If we want to establish a substantial similarity between **I** and **I'**, a formal analogy of both must be supplemented by their material characterization. I will compare deontic, axiological-moral (briefly, axiological) and epistemic modals. Susan Haack (see Haack 2001) distinguished five possible solutions of the comparison problem for epistemic and axiological modals. I will slightly generalize her approach. Let us think about **I** and **I'** as referring to qualifications in terms of deontic, axiological or epistemic modalities, for example, '*x* is obligatory (good, known)'. We have the following relations: (i) **I** is a kind of **I'** (or reversely), (ii) **I** and **I'** are entirely, that is, positively and negatively correlated; (iii) **I** and **I'** are partially correlated (overlap); (iv) **I** and **I'** are mutually not relevant; (v) **I** and **I'** are analogical. We can neglect (v) for the reasons given above. Thus, (i) – (iv). In order to suggest a minimal substantiality, I will consider deontic modals as legal for a while, axiological modals as based on a substantial definition of goodness (it is not necessary to propose how to do that in the present context; one can think about axiological qualifications of legal orders by natural law) and epistemic as defined by (**) or similar conditions for asserting, supposing, assuming, guessing, etc.

The discussed issue has clear links with so-called virtue epistemology and the ethics of belief (see Code 1987, Agazzi 2004, Steup 2006, Greco, Turri 2011 for general surveys; the first book presents a panorama of relations between morality and science). This approach claims that specific epistemic virtues regulate our cognitive acts, that is, make them deontically (in general sense) qualified. For instance, we ought to assert a proposition according to some available or accessible evidence or should abstain from assertion, when we have no sufficient support. According to virtue epistemology, the phrases like 'ought to assert', 'do not assert' or 'to abstain from assertion' express epistemic duties, which can be positive (ought to assert), negative (do not assert) or recommendations to remain neutral (to abstain from assertions). Correspondingly, epistemic rights or possibilities may be also considered, because we can ask when guessing, assuming, supposing, claiming, etc. is permitted. Now, the problem of the relation between the order of epistemic qualifications and the order of moral qualifications appears; Haack uses the term 'evaluation' and I will follow her façon de parler. The view (i) subsumes the epistemic evaluation under the moral one (this is a typical standpoint of the ethics of belief in the strong sense), the position (ii) distinguishes both orders, but simultaneously correlates them (this idea can also be accepted by the ethics of belief), the standpoint (iii) sees both orders as partially overlapping (a weaker form of the ethics of belief can share this view), and the solution (iv) separates both orders. Since I have no business in solving the question which solution is correct, I consider these possibilities merely as an illustration of various eventualities. The comparison of normative modals with axiological ones is obvious. Let me only mention that natural law theorists propose (i) (the strong version) and (iii) (the weak version). On the other hand, legal positivism defends (iv), but can also accept (iv) providing that parallelism of qualifications is only partial and does not influence the problem of validity of the legal norms. The naturalness of comparing deontic (legal) and moral qualification stems from the fact that an axiological order can be considered as a hidden normative system.

It is tempting to compare deontic qualification and epistemic qualification with respect to their normativity. This problem is usually formulated by saying that the concept of knowledge is normative or, speaking more traditionally, that epistemology as such is normative, because it looks for conditions of the validity of knowledge. Of course, we must resign from the earlier assumption that deontic modals are equated with legal ones. Epistemic normativity must mean something different than legal normativity, when we say that epistemology is normative. Yet,

it is also implausible to equate normative epistemic qualification with moral evaluation in the above sense. The mentioned phrases, namely 'ought to assert', 'do not assert' or 'to abstain from an assertion' are now understood as expressing epistemic duties, prohibitions or right. However, we encounter serious problems here. First of all, we have no general theory of normativity. Proposals aiming at such a theory (see Hansson 2001, Wedgwood 2007, Thompson 2008) are so different that considering the concept of normativity as generic, with legal and epistemic (and perhaps other) modals being its species, seems to be quite unrealistic, also due to their commitments to such different general views as cognitivism, non-cognitivism, expressivism, realism, anti-realism, objectivism or subjectivism. On the other hand, there are some reasons to ascribe a kind of normativity to epistemic modals. A preliminary intuition is as follows. Alethic modalities are not normative. Hume's principles, that is, assertions numbered as (14)–(21), do not hold for alethic modals, but they formally characterize normative modalities. Assume that Hume's principles constitute characteristic theses separating the normative from the non-normative. Since epistemic modalities fall under (14)–(21), we have a motive to consider epistemic modals as normative in a sense.

Clearly, the mentioned deontic phrases attributing a normative status to epistemic attitudes appear as conceptually legitimate and not necessarily refer to hidden axiological (moral) evaluations. Moreover, for instance, the statement 'I ought to assert *A*' is not equivalent to 'I assert that *A*'. What does it mean 'ought to' in this context? There are at least two explanatory proposals in this respect. One is proposed by Ralph Wedgwood (see Wedgwood 2007, pp. 153–174) who argues that intentionality is normative. Hence, we can conclude (Wedgwood does not enter this question) that since epistemic states are intentional, they are normative as well. I consider his argument as doubly defective. Firstly, perception is an intentional epistemic (cognitive) state, but its normativity appears to be very problematic. Secondly, Wedgwood seems to confuse intentionality as aboutness with intending in a teleological sense. I see no reason to regard being about as normative, but if we consider intending as genuinely normative, it requires an additional explanation. The second popular account of the normativity of knowledge (or some other cognitive states) assumes that 'I know' is a performative verb (see White 1982, pp. 104–107). Consequently, ascriptions of knowledge function as based on felicity conditions, but they do not appeal to truth-conditions. This suggestion is interesting, but it does not require abandoning of regarding 'I know that *A*' as neither true nor false (it is even more convincing in the case of third-person statements). If we inspect (**) as well as the traditional classical definition of knowledge, we easily find the point in which normativity is hidden. It is exactly the point (**)(c) that a known *A* should be (must be, ought to be) properly (sufficiently) justified. Thus, the normativity of knowledge arises as a result of the normativity of justification. For example, we can say that I should assert *A* if and only if the support for *A* is stronger than data supporting its denial or I can epistemically choose between *A* and $\neg A$ if their support is more or less equal, etc. In general, epistemic attitudes become correct or incorrect with respect to sets of data confirming or disconfirming them. Their correctness is settled by some constraints, which can be termed as normative conditions. Such conditions are sometimes very weak. For example, if I have no data concerning *A*, I can guess or speculate that *A* (or whether *A*), providing that *A* is internally consistent. One warning should be made in this context. Although the normativity of epistemic states appears as obvious due to the normative treatment of justification, details of (**)(c) and similar claims are far from being uncontroversial.

Pure logic provides the only sufficiently clear example of the normativity of justification. Assume that $A \vdash B$, that is, *B* is provable from the assumption *A*. By the deduction theorem, we obtain (##) $A \vdash B$ if and only if $\vdash A \Rightarrow B$. In general, it *A* is a logical tautology of first-order logic, $\vdash A$ if and only if *A* is tautological. Otherwise speaking, for any tautology, *A* and '*A* is tautological' are equivalent. This assertion directly follows from the completeness theorem for first-order logic. This property is nice, because it equates correct assertivity of logical theorems with their tautologicity, that is, universal validity. Since the symbol $\vdash$ is related to rules of inference, a kind of normativity is hidden in it. At the same time, however, the predicate 'is tautological' refers to truth-conditions of logical theorems. Thus, (##) establishes the parity between the logical normativity and the logical indicativity, although 'it is tautological' gives another interpretation of the box. Moreover, (##) is stronger than Tarski's biconditionals, because it holds without any restriction concerning logic, but the universality of **V**-equivalences assumes the classical system. However, there is a high price for (##). It consists in the fact that logical theorems do not distinguish any extralogical content. On the contrary, the contents going beyond logic inevitably cause troubles in understanding what the phrase 'is sufficiently justified' means. A fortiori, the issue of the normativity of justification is notoriously problematic.

VIII
Theology and Logic

Is logic applicable in theology? If so, is logic basically relevant for theology or the former has merely a secondary importance for the latter? Answers to these questions require several preliminary explanations concerning both logic and theology. Let me begin with the *scientia divina.* First of all, there are many theologies, not only because many religions exist. Even if we restrict our attention to three great monotheistic religions (Islam, Judaism, Christianity), we do not find the uniform idea of theology. The situation does not change very much, if we made a further restriction, for example, to Christianity only. The theological tradition of this confession does not appear as uniform. On the contrary, we encounter various conceptions of theology. Two of them are particularly relevant for my further considerations. Firstly, we have the negative or apophatic theology as represented by Nicolaus Cusanus (Nicholas of Cusa), for example. This kind of theology, very popular in the Orthodox Church, says that we cannot say anything positive about God and His attributes. We should abstain from positive assertions and limit ourselves to statements like 'I do not know what God is like' or 'God is not…'. According to this kind of theology, the cognitive gap stemming from such assertions is filled by beliefs as faith. If we believe, we do not need to bother with apparent inconsistencies in the body of theological statements. As Cusanus said, *docta ignorantia* is sufficient and *coicidentia oppositorum* are to be accepted. Clearly, logic plays no essential role in negative theology, which is not particularly interested in arguments.

Secondly, there is theology, which is called natural, rational or philosophical (this last label is not quite correct, because negative theology has an explicit philosophical dimension). This version of the *scientia divina* (see Flint, Rea 2009 for a comprehensive survey) admits and recommends logical arguments in theological debates. Although its particular representatives, for example, Augustine of Hippo, Anselm of Canterbury, Thomas Aquinas, Duns Scotus or Leibniz accept the formula *fides quarens intellectum*, yet their considerably differ as far as the matter concerns the scope of logic as a tool of theology. The standpoint of the official catholic theology, confirmed by decisions of several councils, follows Thomas Aquinas, who distinguished theological truths (for example, the existence of God) accessible to human reason and subjected to rational or logical argumentations and theological truths, which exceed human understanding and are grounded in the Revelation (for example, the doctrine of *creatio ex nihilo*). Moreover, there can be no inconsistency between these kinds of truths, because both are given by God. Thomas' picture, for its compromising character, provides a very good point of orientation. Augustine ascribed more to faith than Thomas did, Anselm and Scotus were more rationalistic than *Doctor Angelicus*, but Leibniz and the Deists of the 18th century became the extreme rationalists in theology.

The rational *scientia divina* forms the proper environment for investigating the role of logic in theology. But what is logic? The noun 'logic' co-occurs with various and mutually different qualifications. Omitting several metaphorical denominations, like "the logic of history" or "the logic of politics", we read about formal logic, transcendental logic, hermeneutic logic, dialectical logic, material logic or philosophical logic. My further considerations are directed to formal logic, because I think that it is the only legitimate kind of logic. We can eventually distinguish logic in the narrow sense and logic in the broad sense. The latter covers semantics or semiotics (roughly speaking, logical theory of language), formal logic and the methodology of science, but the former restrict the scope of logic to its formal part. Since I will touch semantic, formal-logical and methodological problems, it is convenient to work with logic in the broad sense. However, formal logic is the core of logic in any legitimate sense, because it provides the basic criterion of logical correctness. Assume that $A_1, \ldots, A_n, A$ is an argument with $A_1, \ldots, A_n$ as its premises and A as its conclusion. We say that this argument is formally correct or valid, when A logically follows from the set of premises consisting of $A_1, \ldots, A_n$ (symbolically $A_1, \ldots, A_n \vdash A$), but it is materially correct, when its premises $A_1, \ldots, A_n$ are true. This definition entails that if $A_1, \ldots, A_n \vdash A$ and the premises are true, the conclusion must be true as well. This statement exhibits the main property of deduction, namely its soundness or truth-preserving. In general, formal logic is a collection of systems consisting of principles being special cases of logical entailment. Of course, we can try weaken the concept of logical validity and admit partial or inductive correctness, but it has to be done very carefully. In particular, clear features of generalized validity are required. In fact, logic called dialectical, material or transcendental completely lacks, at least until now, precise criteria of the validity of its arguments.

I include logical analysis of various problems into the scope of logic. Logical analysis has a special importance for philosophy. In my view, rational theology is more a kind of philosophy than the *scientia divina* in the traditional sense; this opinion applies to some parts of theology at least, but I will not enter into details. Consequently, natural theology is comparable to philosophy with respect to their status as academic disciplines. This statement has very far-reaching consequences, because if a field has academic character, it has to fulfil several methodological requirements. Thus, applications of logic to theology should be considered as analogous to those performed in general philosophy, if one is inclined to agree that both are similar. I do not claim that logical analysis exhausts the entire scope of philosophy or (*per analogiam*) theology. Yet, several theological problems are suitable for logical treatment. Anyway, the phrase "Logical analysis in theology" could be another title of this paper. Some authors, like Jan Salamucha or Joseph M. Bocheński, Polish theologians-philosophers, claimed that rigorous logical analysis is indispensable for modernized theology (see Woleński 2003 for details of this movement), other were less ambitious, but there is no doubt that the place of logic in rational theology is important. Thus, I do not make any programmatic postulates, but I assert the existing state of art.

I am fully aware that there are several specific problems in theology, which should be taken into account, when methods of theology become a discussed topic. The typical list of problems included into a survey of rational philosophy (see Flint, Rea 2009, p. 5) covers the following problems: (i) Theological Prolegomena; (ii) Divine Attributes; (iii) God and Creation; (iv) Topics in Christian Philosophical Theology; (v) Non-Christian Philosophical Theology. Logical analysis can be used in dealing with any topic taken from the above list. The point (i) has a special character. It concerns (among others) the authority of the Scripture, tradition and the Church, and Revelation and inspiration. Although particular Christian confessions differ in their relation to the Scripture, Revelation, Tradition and Church as authorities, we can assume for the sake of argument that there is the ultimate core of theological opinions. They concern further problems, like divine attributes, God and creation, and several other theological questions. Faithful Christians are obliged to accept theological preliminaries and their consequences for special issues. A secular thinker can complain this situation and say that theological limitations are at odds with the rationality of arguments, particularly, because these restrictions illegally mix natural and supra-natural factors. On the other hand, a Christian thinker might argue that if someone does not accept theological preliminaries, he or she unjustly excludes supra-natural entities from the scope of sound argumentations. This controversy results in the conclusion that any rational exchange of views between believers and unbelievers. I do not agree with this diagnosis. First of all, it appears as unclear why the line of division separates believers and unbelievers. In fact, although various dissimilarities between Catholics and some Protestants as far as matter concerns theological preliminaries force completely different views concerning divine issues, but this fact does not exclude discussions between rival Christian confessions.

I would like to propose a debate without special assumptions, in particular, accusing the opposite side for its irrationality. In order to give a concrete example, I employ a well-known metalogical fact. Assume that **X** is a consistent set of sentences. By the Gödel-Malcev completeness theorem, **X** has a model in which its elements (sentences belonging to it) are true. We do not need to think that these sentences are true in the real world. The model of **X** is a mathematical entity, philosophically speaking, an intentional object correlated with this set by the content of propositions in question. Assume now that these sentences are about God and follow theological preliminaries. According to our assumption, we initially consider **X** as a consistent set of sentences. This assumption can be correct or not and one of possible uses of logic in theology can consist in checking, which answer should be accepted. Even if both parties of the debate, that is, believers and unbelievers, propose conflicting answers to the problem of consistency of theological discourse, they can perfectly understand each other. Similarly, we can investigate several other questions, for example, the problem of the form of propositions about God, the logical status of the term 'God' or the relation between naturalistic and supra-naturalistic language. Step by step, our theo-logic, in order to use a convenient label, might be extended to include discussions about arguments invented to demonstrate the existence of God (are they proofs or persuasive strategies only?) or the relation of God to evil (do attributes of God cohere with the existence of God?). A skeptic might observe that such debates are dramatically inconclusive. Well, but this remark concerns every philosophical topic logically analysed. Thus, philosophical theology is not an exception in this respect and the opposite situation would be surprising. Similarly, an argument that some theological problems – for instance, the Resurrection or the Incarnation – exceed tools of logical analysis does not look convincing, because this is typical for philosophers to maintain that some topics transcend given methods. My position is rather modest and consists in a claim that logical analysis can help in clarifying own positions, defending them and arguing against opposite views. I will give some examples of such analytic arguments. And one additional remark is in order. I think that any participant of a discussion concerning theology should openly declare his or her own relation to religion. I consider myself as an unbeliever (this is a temporary denomination). This attitude determines to some extent that I am interested in the place of logic in arguments for or against God's existence (I this problem as the most central question of religion and theology).

The first problem related to the model-theoretic account of theological discourse pertains to the status of the term 'God'. Assume that we work in the standard first-order logic. If we take the sentence (*) 'God exists', its subject-words look like a proper name (individual constant). Now, first-order logic excludes empty proper names. This is the reason that some believers say that unbelievers cannot discuss the main theological issue. However, it is not true. The unbeliever has no reason to maintain a priori that theology is inconsistent, He or she can, without violating his or her own convictions, accept models in which (a) is true, that is, God exists. Yet, it must be emphasized that to exist in a model is not to exist in reality. If an individual believer and his or her opponent agree at this point, they can argue for their views. We have another device enabling us to achieve the same task. In fact, the logical form of (*) exceeds beyond the syntax of first order-logic, because the verb 'exists' functions as a predicate. In order to work with a form coherent with first-order syntax, we can employ

(**) $\exists x(x = \text{God})$.

in which the word "God" serves as an individual constant, but the existential mark is hidden in the quantificational prefix. Yet (**) assumes that the term 'God' is non-empty. This situation seems to be problematic for some unbelievers, because they deny that God exists. Fortunately, there is a very simple manoeuvre in order to read (**) in a way which satisfies all parties. We can change (**) into

(***) $\exists x(x \text{ is God})$.

The word 'God' occurs in (***) as part of the predicate 'is God'. This predicate has a set as its denotation. Since denotations of predicates can be empty sets, the difficulty noted above disappears. Denote the reference of 'is God' by **G**. Believers maintain that **G** is a non-empty set, but unbelievers think that $\mathbf{G} = \varnothing$. Obviously, the believer should justify that there is an x such that $x \in \mathbf{G}$, but the unbeliever ought to argue that there is no x such that $x \in \mathbf{G}$.

The division into believers and unbelievers is too simple to cover all religious (theological) possibilities. Some people declare that they know that God exists, other say that they know the opposite. The statements 'I believe that God exists' and its negation 'I do not believe that God exists' also render attitudes toward the existence of God. Notorious ambiguity of 'I know' and 'I believe' causes considerable difficulties in establishing a map of possible positions in religious matters. Ordinarily speaking, any motivated belief can be considered as a piece of knowledge. Motives for believing can vary from astrology and superstitions to mature arguments taken from science. They are psychological, cultural, practical, political, etc. This approach to knowledge makes it indistinguishable from merely having beliefs. On the other hand, we have the definition of knowledge as justified true belief. For many reasons (see the Gettier problem, for example), it claims too much. Let us take scientific knowledge as a paradigmatic case of knowledge. We can say that I know that *A* if and only if (a) I believe that *A*; (b) *A* is true according to the present shape of science; (c) I am ready to defend *A* using arguments taken from science. The point (a) treats belief as a mental state, the point (b) suggests that truth is approximate, not ultimate, but the point (c) treats defending as something dispositional. I assume that common knowledge about empirical matters can be regarded as satisfying the points (a) – (c) and call both species of knowledge as cognitively rational. The last qualification has nothing in common with evaluation, positive or negative. Rational knowledge in the outlined sense can be much less valuable from a practical point of view. Rational cognitive knowledge is to be contrasted with beliefs based on the act of faith. Religious beliefs are paradigmatic in this respect. Some believers say that they believe in God and they do not need any rational arguments for their faith. I will consider beliefs based on faith and knowledge (or cognitive beliefs, if you like) as ideal types. Actual mental states are usually a mixture of various and different factors.

I propose a map of attitude toward religion (I follow Woleński 2009. It consists of theism, fideism, atheism and agnosticism as attitudes related to the statement (***) (or (*) or (*). Theism and fideism are frequently considered as equivalent (or almost equivalent) standpoints and the same concerns atheism and agnosticism. My analysis proposes to distinguish theism and fideism on the one hand as well as atheism and agnosticism on the other hand. I guess that the distinction proposed below might help in clarifying discussions between philosophers and ordinary people affirming or denying the existence of God. The theist in Aquinas' style considers (***) as a piece of knowledge, that is, the statement satisfying (a) – (c). Hence, the theist of this kind can express his position by

(1) I know that God exists.

What should the atheist say? If he or she replies that God does not exist, the theist might ask "Well, but do you know that?" The atheist has a problem, because if his or her position is considered as the negation of theism, the answer should be expressed by

(2) I do not know that God exists,

but this is too weak. The atheist is rather inclined to say

(3) I know that God does not exist.

Now (2) and (3) are not mutually contradictory, but contrary. This simple reasoning shows that there are other possibilities in attitudes toward (***) then (1) and (2), if the latter is taken as an adequate expression of atheism. The matter is still more complicated if we observe that

(4) Although I do not know that God exists, I believe that He exists.

Since many historical fideists accepted (4), this possibility is not abstract. Thus, a closer analysis of theism, atheism, etc. is required. In particular, we should take into account knowledge and belief as parameters important in attitudes related to (***). How are they related? I accept the following dependence (KA – I know that A, BA – I believe that A)

(5) $KA \Rightarrow BA$,

This formula is not a logical axiom but rather a result of an analysis of knowledge and belief. It can be justified by the account of knowledge as true justified belief. Even if we consider "true" and "justified" in the above definition of knowledge as somehow problematic, (5) can be taken as fully granted for any kind of belief, including religious faith. The reverse dependence, that is, $BA \Rightarrow KA$, does not hold. Speaking metaphorically, belief and knowledge are half-separated: although knowledge implies belief, the latter does not imply the former.

In order to go further, I will establish some formal properties of K and B. I repeat the diagram (**D2**) from Chapter VIII as instrumental in this respect:

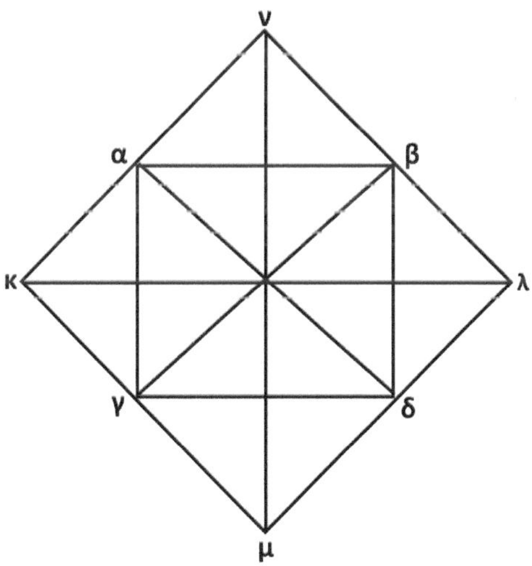

This diagram is a generalization of the well-known logical square for categorical and modal propositions. Let symbol ■ denotes K or B. Understanding α as ■A, β as ¬A, γ as ♦A and δ as ♦¬A, we have the following facts (among others):

(6)	$\neg(\alpha \wedge \beta)$	(α and β are contrary);
(7)	$(\alpha \Rightarrow \gamma)$	(α entails γ; γ is subordinated to α);
(8)	$(\beta \Rightarrow \delta)$	(β entails δ; δ is subordinated to β);
(9)	$(\alpha \Leftrightarrow \neg\delta)$	(α and δ are contradictory);
(10)	$(\beta \Leftrightarrow \neg\gamma)$	(β and γ are contradictory);
(11)	$(\gamma \vee \delta)$	(γ and δ are complementary);
(12)	$(\blacksquare A \Leftrightarrow \neg\blacklozenge\neg A)$	(■ is definable as $\neg\blacklozenge\neg A$);
(13)	$(\blacklozenge A \Leftrightarrow \neg\blacksquare\neg A)$	(♦ is definable as $\neg\blacksquare\neg$)
(14)	$(\beta \Leftrightarrow \blacksquare\neg A)$	(β is definable as $\blacksquare\neg A$);
(15)	$\nu \Leftrightarrow (\alpha \vee \beta)$	$(\nu \Leftrightarrow (\blacksquare A \vee \blacksquare B)$);
(16)	$\mu \Leftrightarrow (\gamma \wedge \delta)$	(16) $\mu \Leftrightarrow (\blacklozenge A \wedge \blacklozenge\neg A)$;
(17)	$\neg(\nu \Leftrightarrow \mu)$	(ν and μ are contradictory);
(18)	$\alpha \vee \beta \vee \mu$	(α, β, μ are mutually exclusive and jointly exhaustive).
(19)	$\kappa \Leftrightarrow A$;	
(20)	$\lambda \Leftrightarrow \neg A)$;	
(21)	$\neg(\alpha \Rightarrow \kappa)$	$(\neg(\blacksquare A \Rightarrow A))$;
(22)	$\neg(\kappa \Rightarrow \alpha)$	$(\neg(A \Rightarrow \blacksquare A))$;
(23)	$\neg(\lambda \Rightarrow \beta)$	$(\neg(\blacksquare\neg A \Rightarrow \neg A))$;
(24)	$\neg(\beta \Rightarrow \lambda)$	$(\neg(\blacksquare\neg A \Rightarrow \neg A))$.

These dependencies – in particular, (21) – (24) – normalize the operators K and B as functioning in non-normal modal logic. It is important to observe that knowledge that A does not imply that A is true, but this fact is not at odds with (a) – (c) (see above). The main reason is that 'to know' and 'to believe' are strongly intensional words. A well-known criterion of the sentence $I\exists xPx$ implies neither $\exists xPx$ nor $\neg\exists xPx$, if the letter I stands for a strong intensional operator. By analogy, neither KA nor BA implies A. Note also that (5) has no justification in (D). However, we can say that accepting $KA \wedge \neg BA$ would not be rational, because if someone knows that A, he or she should (it is cognitively rational to) believe that A as well. Of course, there are always problematic cases of assertions like 'I know that A, but I do not believe that A' (this example puzzled Moore and Carnap), but if we distinguish consistency and rationality, (5) is defensible via an appeal to the latter. Some special views concerning the plausibility of $BA \Rightarrow KA$, in some cases strengthened to the equivalence $KA \Leftrightarrow BA$, will be discussed later.

If we accept (5) as a conceptual truth or an additional axiom, we have the following consistent possibilities:

(25) $KA \wedge BA$;

(26) $\neg KA \wedge BA$;
(27) $\neg KA \wedge \neg BA$.

Thus, I can know that *A* and believe that *A* or not know that *A* and believe that *A* or not know that *A* and not believe that *A*. Further, not knowing that *A* is consistent with believing that *A* as well as not believing that *A*, but not with both simultaneously. Consequently, not knowing that *A* does not imply anything definite about believing. Some additional possibilities appear when we consider

(28) $K\neg A \Rightarrow B\neg A$.

This generates the following coherent cases

(29) $K\neg A \wedge B\neg A$;
(30) $\neg K\neg A \wedge B\neg A$;
(31) $\neg K\neg A \wedge \neg B\neg A$.

By combining, (25) – (27) and (29) – (31), we obtain:

(32) $\neg K\neg A \wedge \neg KA \wedge BA$;
(33) $\neg K\neg A \wedge \neg KA \wedge B\neg A$;
(34) $\neg KA \wedge \neg BA \wedge \neg B\neg A$;
(35) $\neg K\neg A \wedge \neg BA \wedge \neg B\neg A$;
(36) $\neg KA \wedge \neg K\neg A \wedge \neg BA \wedge \neg B\neg A$.

Now, if we skip formulas which are proper parts of others or follow from others, there remain (25), (29), (32), (33) and (36).

If we apply the above considerations to (*), we obtain:

(37) I know that God exists and I believe that God exists;
(38) I know that God does not exist and I believe that God does not exist;
(39) I do not know that God exists and I do not know that God does not exist and I believe that God exists;
(40) I do not know that God exists and I do not know that God does not exist and I believe that God does not exist;
(41) I do not know that God exists and I do not know that God does not exist and I do not believe that God exists and I do not believe that God does not exist.

Combination (37) expresses the standpoint of the Catholic Church on which knowledge and faith (religious belief) are complementary and mutually irreducible; the famous formula *fides quaerens intellectum* displays this view. (38) is characteristic of atheism popular in the French Enlightenment or in Sartre. (39) is the *credo* of fideism. (40) defines agnosticism and (41) describes the radical epistemological-theological skepticism (I will skip this view, because I consider it as not interesting very much). Theism and fideism contain the statement 'I believe that God exists' as their common part and this statement can be taken as expressing theism (or fideism) in a wider sense. Similarly, atheism, agnosticism and radical skepticism, as given by (41), agree with respect to the assertion 'I do not believe that God exists', which characterizes atheism in a broader sense; this picture is historically faithful, because all three views were and are pointed out as atheistic. Atheism and agnosticism accept the statement 'I do not know that God exists and I believe that God does not exist'. I propose to count this attitude as indirect atheism. If we take the set {theism, fideism, atheism, agnosticism}, all standpoints listed in it are mutually independent and contrary, but not all are mutually contradictory. Theism in the broader sense and indirect atheism are also contraries, but the former is inconsistent with atheism in the broader sense.

The above analysis is based on a purely formal criterion. This does not preclude that particular views differ in their account of how knowledge and faith are related. Now I can supplement my remarks (see above) about various motives of religious belief. Let me start with fideism. The fideist will probably say that he or she accepts (5), but provided that knowledge as applied to religion is of a special kind and, therefore, 'I know' in the religious discourse has a specific meaning, different than in the statement 'I know that 2 is the first prime number'. Thus, when the fideist announces that we have no rational knowledge of God's existence, that is, based on intersubjectively accessible empirical evidence, this does not exclude another kind of experience, for example, mystical experience. A more ordinary understanding of fideism is perhaps given by

(42) I believe that God exists, although I have no rational support for my belief and I claim that there are no rational reasons for denying God's existence.

Yet, I think that (39) gives a better account of fideism, at least for its logical analysis and provided that 'I know' refers to rational knowledge. Logical reasons also dictate that I prefer 'I know (do not know) that God exists' than 'I know (I do not know) whether God exists'. The theist of the Thomistic brand (or more generally, recommended by the Catholic Church) accepts the equivalence (i) $KA \Leftrightarrow BA$ and justifies this by the above quoted formula *fides quaerens intellectum*, at least insofar as the matter concerns the existence of God. Pragmatically speaking, faith (religious belief) is the starting point and its equivalence with knowledge is dictated by *ratio recta.* Since (i) holds for $\neg KA$ and $\neg BA$ as well, the theist has to accept it too. According to standard theism, (ii) $\neg KA \Leftrightarrow \neg BA$ describes unbelievers. They do not believe if and only if they do not know. Their lack of knowledge and belief may be unconscious and non-culpable, but if it is deliberate, they are deprived of *ratio recta*. Paradoxically, the atheist also shares (i) and (ii), but the motivation is different here. First of all, the atheist strengthens (ii) to (iii) $K\neg A \Leftrightarrow B\neg A$, because atheism consists in a positive view about God's non-existence, not only in expressing the lack of knowledge and faith concerning his existence. Thus, the atheist asserts (iv) $K\neg A \Leftrightarrow \neg KA$, (v) $B\neg A \Leftrightarrow \neg BA$, and (vi) $K\neg A \Leftrightarrow B\neg A$, and argues that the theist abuses the concept of knowledge. In other words, the atheist maintains that belief and knowledge as attitudes concerning (***) should be entirely rational. This position is best expressed by (vi), but (ii) holds only because *KA* and *BA* are false. Finally, the atheist can agree that *KA* and *BA* are true, but knowledge, belief and truth are understood differently than in rational discourse. Since (i) – (vi) do not follow from (D), the rationale behind their acceptance or rejection has its source in decisions motivated by various factors.

The agnostic proceeds similarly to the fideist in the matter of the relation between knowledge and faith, namely – he or she derives some consequences from the lack of knowledge. However, agnosticism insists that knowledge should be taken as rational: belief is rational if it is grounded by knowledge – otherwise, is irrational. Thus, there are obvious affinities between atheism and agnosticism. The skeptic, according to his or her general epistemological position, is completely neutral about the above issues. Skepticism is not particularly interesting, because it consists of merely negative assertions. The remaining views propose at least some positive beliefs, but theists and atheists additionally insist that their adherents know something about God's existence or non-existence. A positive element in belief or knowledge seems to be of the utmost importance for world-view matters, because negative declarations do not suffice in this area. Although, on the one hand, every view (I omit skepticism) in the set {theism, fideism, atheism, agnosticism} is contrary to the rest; on the other hand, theism and atheism operate on a different epistemological level than fideism and agnosticism. In particular, the latter assume epistemological skepticism concerning the knowledge of God's existence, but reject theological skepticism in this question. This positive content, directly related to theology in the case of fideism and indirectly in the case of agnosticism, is contained in statements 'I believe that God exists' and 'I do not believe that God exists', respectively. Similarly as in the case of theism and atheism, both the agnostic and fideist approaches to (i) – (vi) exceed the logical space. If we start with 'I do not know whether God exists', we have two choices concerning God's existence, namely – we can either assert or deny it; fideism chooses the first possibility, agnosticism opts for the second one.

Saying that decisions in world-view matters are practical and motivated by many factors, I do not deny that defenders of various attitudes towards religion support their own positions and criticize those of their opponents. Yet, due to the variety of evidence involved in the proposed arguments, we have an actual mosaic of issues and their epistemological qualifications. Let me appeal to historical examples. The theism of the Aquinas and similar thinkers looks for deductive arguments for God's existence, eventually supplemented by data from physics and metaphysics. Anselm of Canterbury and other theological rationalists formulate ontological proofs, which are deductive and entirely conceptual. Circumstantial proofs or arguments for God's existence invoke very different circumstances. For example, the defenders of the idea of Intelligent Design frequently say that the mathematical nature of physical reality cannot be explained without an appeal to God. Other defenders of religion point out miracles, historical facts (for instance, Jesus' life), morality or mystical states. It is difficult to say something more about this variety than to point out that its elements have a very different argumentative force. The situation is additionally complicated by the fact, confirmed by almost every world-view discussion, that the use of arguments pro and contra (*) is very deeply determined by the positions of people involved in such debates, much more than in scientific or even philosophical controversies.

Nevertheless, the difficulties in assessing and collecting possible arguments for God's existence or against this belief do not preclude some general observations about the strategies of theism, fideism, atheism and agnosticism; I exclude skepticism once again. In particular, we can say something about *onus probandi* in related proofs or disproofs. The *onus probandi* (the burden of proof) indicates which part in a discussion should prove or justify its thesis. In general, the *onus probandi* belongs to the party which asserts something positively. Clearly, the issue primarily concerns (***) (this form is the most suitable for a logical discussion), because proving this sentence motivates the theist to take his or her position captured by (37). Hence, the atheist is not obliged to prove the negation of (***). Thus, the demand, frequently directed to the atheists, that they should prove God's non-existence, is simply misguided for logical reasons. This does not mean that the atheist has nothing to do with justifying his or her own standpoint. In general, the atheist asserts that the set {God} is empty. One argument consists in objecting to given proofs that this set is non-empty. This way is of limited force, just because the theist can always say that other proofs are possible. Other atheistic arguments try to show that no empirical data suggest that the predicate "is God" refers to a non-empty set. For example, many atheists claim that no empirical support can be given for assertions about beings devoid of empirically graspable arguments. Further analysis leads to a substantial discussion concerning reasons invoked by defenders of theism and atheism. Since the fideist does not appeal to knowledge (e.g. scientific knowledge) of God or His existence, he or she limits his or her arguments to personal ones. Thus, fideists consider the *onus probandi* as related to 'I believe', not to 'I know'. Similarly, the agnostic stresses 'I believe that God does not exist' and claims that there are no sufficient reasons for believing that God exists. In relation to God, fideists and agnostics, contrary to theists and atheists, are more interested in believing than in knowing. They agree that knowledge (in the rational or scientific sense) does not apply to God, but they can disagree whether there are other kinds of knowledge or what the word 'knowledge' means.

Historically speaking, atheism and agnosticism usually point out the lack of scientific knowledge of God and insist that other types of knowledge are merely personal and subjective cognitive states and thereby cannot pretend to objectivity. This last observation well shows how the whole issue is complicated and what are the consequences of the fact that theism and atheism operate on different epistemological levels, although fideism is closer to theism and agnosticism is closer to atheism, in particular, if we take into account practical consequences of the discussed views. There is also a psychological argument which applies to both to atheists and agnostics. Let us call them disbelievers for a while. Religious attitudes, whether understood as knowledge or merely as faith, have their grounds in positive mental states. Of course, one can say that disbelievers are deprived of such states and that the absence of a mental state is also a mental event. That is true, but there are serious doubts as to whether a negative state of affairs can be causally efficient. However, the practice of religious disbelievers suggests that, if their atheism or agnosticism is not caused by growing in a non-religious environment, they always look for weak points of theological arguments.

Demonstratio appeared in mathematical texts in the 17th century with regard to speaking about deductive proofs (in our sense). The letters QED (*quod erat demonstrandum*, 'which was to be proved') still remind of this terminological custom. It is interesting that the word *deductio* had a different meaning at that time. Its employment in medieval Latin was related to law and applied to deriving a claim or right before a court. The process of such derivation was divided into two parts, namely factual and legal demonstration. Assume that a person *Y* borrowed money from a person *X* and did not return the loan. If *X* put the case before the court, he had to prove that the loan actually took place and that there is a legal rule, which generated the claim in question. Otherwise speaking, one problem concerned *quid facti?*, and the other concerned *quid juris?* Kant used this sense of *deductio* in explaining his idea of the transcendental deduction, pointing out that every case of this reasoning has a *quid facti?* part and a *quid juris?* part. When Kant was speaking about deduction in our sense, he used the term *Ableitung* (derivation by logically admissible steps). Of course, *Ableitung* is analytic, apodictic and, thereby, a priori. For Kant, the transcendental deduction has the same nature, except that it is analytic. Thus, Kant argued that the most important pattern of reasoning is apodictic and a priori, but it is not reducible to purely logical steps. A less philosophically sophisticated usage occurred in Newton, who also employed *deductio* in a very loose sense as deriving hypotheses on the basis of some evidence, but who proved the theorems in the *Principia Philosophiae Naturalis* by rigorous *demonstratio.*

The development of logic and metamathematics in the 19th and the 20th centuries resulted, among other things, in establishing the meaning of the term 'deductive proof' in the sense of Kant's *Abletung* in the following manner. The deductive proof of a sentence A (more generally, a formula; but we can speak about sentences in the present context without any loss of generality) on the basis of a set **X** of sentences – notation: **Pr**(A, **X**) – is a finite sequence of sentences **S** = $A_1, \ldots, A_{n-1}, A_n$ such that $A_n = A$ and for any $1 \leq i \leq n$, A_i is either a logical axiom (A_i is provable from the empty set, that is, **Pr**(A_i, ∅)) or $A_n \in$ **X** or A_n is provable (derivable) from the earlier items of **S** by applying merely logical rules of inference. By a logical rule of inference we mean a rule which preserves truth, that is, never leads to false conclusions if the premises are true. Accordingly, the concept of deductive proof essentially uses the notion of logical consequence or logical entailment. Thus, we can say that A is a logical consequence of **X** or **X** logically entails A instead of saying that A is provable from **X**. If **X** is an axiomatic system, we speak about axiomatic proofs, that is, proofs based on some sentences that are adopted as axioms. Traditionally, one says that axioms are accepted without proof. However, if one writes axioms, one automatically constructs one-term sequences and any such sequence can be considered as a proof for an axiom which constituted this sequence. In the case of the axiomatic proof (as in geometry), every member of **S** is an axiom of a consequence of axioms obtained by the logical rules of inference. By conditional proofs or proofs from assumptions we understand derivations in which **S** consists of assumptions, not necessarily considered as axioms. If A is an axiom, we think of it as a true sentence, unless it is demonstrated as incorrect. This presumption does not concern conditional proofs, because these assumptions are only accepted, but not accepted as true. For example, we can derive deductive conclusions from falsehoods or sentences temporally or hypothetically accepted. Since the general concept of proof does not require that all elements of **S** are true, one traditionally distinguishes formal and material correctness of proofs. A proof **Pr**(*A*, **X**) is formally correct if *A* is logically entailed by **X**, but it is materially correct if it is formally correct and its premises are true. It is taken for granted that axiomatic proofs are materially correct. This allows us to say that proving *A* consists in proving that *A* is true, but this is an oversimplification. Anyway, the concepts of proof and deductive proof can be considered as equivalent under the above analysis.

Not all arguments are deductive. Consequently, if we accept the equivalence of proofs and deductive proofs, not all arguments fall under the form **Pr**(*A*, **X**). This looks trivial, but it is not, because we should distinguish between worthless non-deductive arguments and good non-deductive arguments. If I say that tomorrow it will rain because one week ago it was Friday, my argument entirely fails, but if I say that tomorrow will snow because today it is cold, this reasoning has some credibility. Now, there are well-known problems with good non-deductive arguments consisting in difficulties in defining non-deductive goodness, soundness, correctness, etc. Let us assume, without attempting to state how the qualities of non-deductive arguments might be measured, that we have intuitive criteria in this respect. The tradition of using such words as *apideixis*, *demonstratio* or 'proof' suggests broadening the concept of truth in order to apply it to good non-deductive arguments. In general, such arguments consist in acquiring empirical evidence to support the proposed statements or hypotheses. Legal arguments, related to the question *quid facti?* and empirical reasoning in science are paradigmatic examples of proofs in the presently considered sense. Let us observe that the word 'proof' has a definite rhetorical flavour. One could propose that we should use the generic term 'justification' and divide justifications into deductive (proofs) and inductive (in the broad sense), eventually the latter into good and worthless. Under such a convention, every proof is a justification, but not reversely. However, it is difficult to expect that this proposal would be accepted, perhaps except by philosophers of science. For example, judges will say "We do not merely justify our statements concerning that someone is guilty, we must prove them". Similarly, when Newton said *hypotheses non fingo*, what he had in mind was to propose something proved, not merely justified. The reason for such an attitude seems to be related to an evaluative or normative aspect of the meaning of 'proof', suggesting that if something is proven it is beyond the pale of reasonable doubt, at least subjectively. This observation decides that *apodeixis* is understood as apodictically convincing (the same concerns *demonstratio*), although etymology does not suggest this. Anyway, according to common opinion it is much better to prove than to merely justify.

Deductive proofs and proofs from various empirical circumstances are closely related. As it has been observed many times, not everything is deductively provable. Even if we agree that logical tautologies are derivable from the empty set and proofs of axioms are one-term sequences, it remains to justify that these, but not other axioms are adopted. The issue is particularly important when postulates of empirical theories are proposed, but also mathematical axioms cannot spring from nothing. I will illustrate this issue by an example relevant for the main topic of my paper (see Sobel 2004 for the comprehensive survey of pros of God's existence). The proofs of the existence of God *ex motu* and *ex causa efficiendi* fall under the following scheme:

(****) if $\forall x \exists y P(x, y)$, then $\exists y \forall x P(x, y)$.

The following arguments are instances of (****): (a) if every moving thing is moved by its mover, then there exists a mover of all things; (b) if every thing is an effect of a cause, then there exists a cause of all things. Clearly, (****) is invalid. As counterexample, we can take a fallacious argument saying that if for every natural number *n* there is a number greater than *n*, then there is the natural number greater than all natural numbers. Russell in his famous conversation with Father Frederick Copleston used another example, namely that the assertion that there exists the mother of all people does not follow from the assumption that every man has the mother, because the former is false and the latter true.

Thomas Aquinas supplemented (****) in his proofs by reasoning to the effect that infinite sequences of movers or efficient causes are impossible. I do not evaluate whether his *demonstrations* are sound or not. He appealed to Aristotelian physics, while his followers are rather inclined to rely on metaphysical necessity. Anyway, Aquinas intended to prove the existence of God. He used the label *demonstratio*, applied syllogistic forms and proceeded by *reductio ad absurdum*. Although his reasoning also appeals to empirical circumstances, the Five Ways are deductive and have the structure of syllogisms. Thus, the interpretations of Aquinas, which point out (see, for example, Schmidt 2001 in the introduction to Part I) that he did not offer proofs, but only suggestions for unbelievers or defenders of the faith, are historically unfaithful. There is still another reason to think that Thomas Aquinas considered the Five Ways as proofs. His idea of religious belief considered every act of faith as veridical, if comes from right reason (*ratio recta*). This means that *ratio recta* produces knowledge, independently of whether it concerns religion or not. There can be a problem with revealed truths that transcending human cognitive capacities (for example, the Trinity Dogma), but (***). This statement is subjected to a normal *demonstratio* by syllogisms and *reductio ad absurdum* employing factual information. Aquinas, a very faithful follower of Aristotle, could not accept knowledge without it being proven. Anyway, nothing prevents one from considering Aquinas' and other proofs as arguments about theological models in the sense exhibited at the beginning of this paper.

The role of considering a theological argument as concerning some model is particularly impressive in the case of ontological proofs of God's existence. I will comment on a proposal made by late Jerzy Perzanowski (see Perzanowski, draft). Perzanowski offers a reconstruction of the reasoning proposed by Charles Harthshorne. It is based on two axioms:

(a) if a being is the most perfect, its existence is necessary;
(b) the existence of the most perfect being is possible (Leibniz' lemma).

Using the modal system **S5**, we can obtain:

(c) the most perfect being exists.

Perzanowski simplified Harthshorne's argument by proving (b) in a stronger modal logic equating truth and necessary truth. Moreover, this logic proves (this is another version of the Leibniz lemma) that

(d) if the most perfect being is possible (its existence is possible), it is necessary (its existence is necessary)

Hence, via modus ponens, we have

(e) the most perfect being exists as necessary (necessary existence implies existence).

The last step ends Perzanowski's reconstruction.

If the matter concerns (a), one might observe that it is a conditional assertion having an existential antecedent. Hence, any further application of this axiom essentially depends on the truth of the sentence 'the most perfect being exists'. If (b) is applied in proving that the most perfect being exists, the entire argument is burdened by *petitio principii.* This is quite evident if we consider the equivalence:

(f) the most perfect being exists if and only if the most perfect being exists and it is possible.

However, (f) is trivial because the sentence '*a* exists' entails 'it is possible that *a* exists'. As far as the issue concerns Perzanowski's simplification, the success of his main move based on (d) requires a very strong modal logic (the logic of strong rationalism). Although the scope of the term 'logic' is conventional to some extent, one can express some serious doubts whether logics should lead to so strong existential consequences. Personally, I would prefer to say that the proof of the Leibniz lemma and further steps of the entire argument proceed in the context of some formal theory involving modal concepts and relations between them and that this theory is not purely logical. Nevertheless, the fact that Anselm's argument can be presented via a precise formal machinery shows the power of formalization applied to concepts of rational theology or theological ontology (see also Tkaczyk 2007 for embedding Anselm's proof into first-order logic without modalities).

Since, according to Leibniz, possibility is logically equivalent to consistency, (d) can be rendered as

(g) if the most perfect being is consistent, its existence is necessary.

Let the letter ***B*** denote the predicate 'is the most perfect being'. Assume that ***B*** is introduced by some consistent set ***K*** of sentences. By the Gödel-Malcev completeness theorem (a set ***X*** of sentences has a model if and only if this set is consistent), ***K*** has a model. This assumption metamathematically guarantees that ***K*** is non-empty. One can be even tempted to say

(h) if a set of sentences has a model, it possesses it necessarily (it is impossible that the set in question has no model).

However, this last assertion requires additional constraints (see below). The way in which ***K*** is given as defining the item conceived as greater than any other being, is very important. First of all, the phrase 'the conceived item' means 'the item described by a set of sentences'. Secondly, according to the common opinion, the most perfect being (**MPB**, for brevity) is introduced by the maximalization procedure. Metalogically speaking, it consists in the identification of **MPB** with a collection of perfections (omniscience, omnipotence, the greatest goodness, immutability, infinity, etc.); existence belongs to perfections. Assume that $\boldsymbol{K^0}$ is an initial consistent collection of perfections. It might be maintained that the application of the Lindenbaum maximalization theorem (every consistent set of sentences has a maximally consistent extension) provides an argument for generating the adequate set ***K***. However, the Lindenbaum extensions are not unique. More specifically, if ***X*** is a consistent set of sentences, it has more than one (in fact, there are infinitely many) maximally consistent extensions. Every such extension ***E*** has a model (a possible world) in which elements of ***E*** are true. Moreover, since these sentences cannot be false in this model, they must be true in it as well. Note that the relativization to a specific model is crucial.

Let us apply these observations to the set $\boldsymbol{K^0}$ and ***K*.** Clearly, the former set has several different maximally consistent extensions, including ***K*** as their part; in fact, ***K*** itself is not maximal, but it is a minor point. Metamathematical observations about ***K*** do not suffice to demonstrate that this set is true in every possible world. Note that we could consider ***K*** as a body of absolute necessities (necessary truths), that is, sentences true in every model. It is obvious now that necessity of truth with respect to a specific model is not absolute, but just relative, because a sentence, which is true in one model, can be false in other possible worlds. Moreover, there is no a priori reason to refer to one and only one initial set $\boldsymbol{K^0}$ of perfections. One could ask, for instance, why the greatest goodness or immutability, but not changeability or goodness directed to people deserving it to some degree, should be regarded as *prima facie* perfections. Typically listed perfections are recommended by Christian (or other similar) theology, but it is only a religious argument, not a logical one. Thus, contemporary versions of Anselm's ontological proof do not show that ***B*** is not empty in virtue of logical necessity. Further, one cannot demonstrate independently of assumptions of a given monotheistic theology that the denotation of ***B*** is unique.

If we apply a similar technique to any consistent collection of anti-perfections (that is, pointing out the lack of perfections in the ordinary or theological sense), it is easily provable by the dual logic and metamathematics (it takes falsity as the distinguished value) that there is such a being that nothing lesser can be conceived. We can interpret this being as the Absolute Evil (**AE**, for brevity). If we entirely omit the ordinary or religious meaning of perfection, **AB** is a maximal being, because it is constructed by a similar maximalization strategy as employed in the case of the denotation of ***B***. There are no reason, at least logical, to maintain that **AE** is less real than **MPB** or Christian God. This leads to a Manichean theology with its perennial battle between forces of Goodness and Evil or even a radicalized Marcionic heresy according to which the world presents itself as an emanation of the personified Evil, in particular, deceiving people in order to make them suffer greater. The theist could presumably answer that the metamathematically phrased ontological proof is enough for demonstrating the existence of **MPB**, that is, say, a god of philosophers, having properties not entirely coherent with Christian theology. This being is necessary and thereby existing. However, this standpoint is not correct, even if we agree that some perfections actually occur in our world. Lindenbaum's maximalization procedure does not imply that there exist maximal perfections in the sense of Anselm and his followers. Although people know something, can do something or behave properly, these facts do not entail that there is a being (subject) which realizes these properties in the most maximal manner. Thus, there are models in which no **MPB** occurs. In fact, some maximally consistent extensions of $\boldsymbol{K^0}$ contain the sentence "there exists the **MPB**, according to a given specification of perfections", but this sentence is false in other extensions. Since we do not know which model represents our universe, we can only say that gods of philosophers (and theologians as well) exist in some possible models (see also Perzanowski 2012, where one can find a good example of model-theoretical thinking in theology; Perzanowski gives a set of definitions for monotheistic philosophical theology).

Although I am skeptical about scientific knowledge that God does not exist, I do not suggest that the atheist has no arguments for justifying his or her own standpoint. In general, the atheist asserts that the set {God} is empty. One argument consists in objecting to given proofs that this set is non-empty. This way is of limited force, just because the theist can always say that other proofs are possible. Other atheistic arguments try to show that no empirical data suggest that the predicate 'is God' refers to a non-empty set. For example, many atheists claim that no empirical support can be given for assertions about beings devoid of empirically graspable arguments. Further analysis leads to a substantial discussion concerning reasons invoked by defenders of theism and atheism. Since the fideist does not appeal to knowledge (e.g. scientific knowledge) of God or His existence, he or she limits his or her arguments to personal ones. Thus, fideists consider the *onus probandi* as related to 'I believe', not to 'I know'. Similarly, the agnostic stresses 'I believe that God does not exist' and claims that there are no sufficient reasons for believing that God exists. In relation to God, both fideists and agnostics, contrary to theists and atheists, are more interested in believing than in knowing. They agree that knowledge as cognitively rational enterprise does not apply to God, but they can disagree whether there are other kinds of knowledge or what the word 'knowledge' means. Historically speaking, atheism and agnosticism usually point out the lack of scientific knowledge of God and insist that other types of knowledge are merely personal and subjective cognitive states and thereby cannot pretend to objectivity.

Agnosticism is my own position. It is motivated by logical difficulties of all known proofs of God's existence. They allow me to reject (***) and, consequently, to accept the statement 'I do not believe that God exists'. Additionally, I argue that the language of theology and the language of science are semiotically incommensurable. Since any proof (deductive or empirical) assumes that the language in which it is embedded possesses semantic uniformity, prospects of proofs from the world to transcendence or in the reverse direction are not very great. On the other hand, logic helps to understand the nature of theological enterprise and its quality.

IX
Ens et Verum Convertuntur (Are Being and Truth Convertible)? A Contemporary Perspective

The theory of transcendentals was developed by the Schoolmen in the Middle Ages and constituted the central part of scholastic ontology, particularly of Thomas Aquinas.[5] According to this theory, transcendentals are the most general concepts. The list of transcendentals varied from author to author, but *ens* (being), *verum* (truth) and *bonum* (goodness) were always included. *Unum* (one), *res* (thing) and *pulchrum* (beauty) were also posited as transcendentals. The concept of being plays a special role among transcendentals, because other concepts are compared to it.[6] The basic dependence holding for transcendentals is captured by

(*) If *T* and *T'* are transcendentals, both are mutually convertible.

Let me note that Duns Scotus proposed a different theory of transcendental concepts (I use the expressions 'transcendentals' and 'transcendental concepts' interchangeably). He distinguished also the so-called disjunctive transcendentals, such as necessity, possibility and contingency, which do not satisfy (*). Clearly, since being can be either necessary or contingent (merely possible), the concepts of necessity and possibility are not the most general ones.

The scholastic theory of transcendentals is essentially rooted in Aristotle's ontology. The Stagirite drew a distinction between primary and secondary substances, that is (in Latin terminology), *entia per se* and *entia in alio.* These can be divided into ten distinct categories. Disregarding various details related to the content of the Aristotelian ontological doctrine, concerning form and matter or matter and energy, individual substances are the only sort of *entia per se*, while properties of primary substances and relations holding between such substances (for instance, quality, quantity, time, position or action) constitute the remaining nine categories. Since being is not a category, it does not belong to the hierarchy

5 For historical accounts, see (in chronological order) Knittermeyer 1920, Schulemann 1929, Wolter, 1946, Elders 1992, Aaertsen 1996, Aertsen 2012.

6 Being is understood here in the distributive sense, i.e. as a collection of objects, not in the collective sense, i.e. as a mereological whole.

of categories. Hence, being and other transcendentals satisfying the principle (*) should be viewed as transcategorial (transcendental) concepts. Note that this understanding of transcendentality differs radically from the Kantian account.

Generally speaking, categories are expressed by universal concepts (universals). Consequently, transcendentals are not universals. This statement entails important consequences. Firstly, transcendentals cannot be defined *per genus proximum et differentiam specificam*, because if *T* is a transcendental, no genus can be associated with it. Secondly, assume that *u* is a universal. Thus, it has its opposite non-*u*, for instance, if a dog (as a concept) is a universal, non-dog serves as its opposite. The universals *u* and non-*u* fall under a more universal scope, let us call it *u'*, and exhaust it; symbolically, *u* + non-*u* = *u'* (the notation is conventional). In our example, dog and non-dog fill the scope of being an animal (supposing that we consider animals as the background set in this case). Assume that *P* is a universal. Thus, non-*P* is also a universal. In general, if *u* and non-*u* are universals, there exists the universal *u'* such that *u'* is a genus, and *u*, non-*u* function as its species. This property allows to define universals by *genus proximum et differentiam specificam*; Consequently, transcendentals do not have their opposites as concepts. In particular, the word 'non-being' does not express any concepts. In medieval logic the operation of forming the opposite concept was called *infinitatio* and consisted in putting 'non' before a noun. However, as far as the transcendental are concerned, *infinitatio* cannot be applied to such concepts. *A contrario*, universals are the only concepts allowing the operation of *infinitatio*; I disregard here the issue of individual concepts. The problem of nothingness was a practical outcome resulting from the absence of the opposites for transcendental. There is no nothing, because 'nothing' is not a concept and cannot refer to any being.

The above version of the theory of transcendentals is still very popular among the Neo-Thomists. On the other hand, its general ontological assumptions are too obsolete for contemporary philosophers. Fortunately, the theory itself can be modernized in the analytic spirit, if only by using tools of the present-day logic and semantics.[7] In particular, (*) take the form

(**) If *T* and *T'* are transcendentals, both are co-extensional.

Clearly, (**) is not enough to characterize transcendentals in a contemporary setting, even if we add that *T* and *T'* are not co-intensional, because we should

---

7 Let me add that some Polish Thomists, notably J. Salamucha and J. M. Bocheński anticipated this way of viewing the theory of transcendentals (see Woleński 2003 for a presentation of their views).

explain the difference in intensions of transcendental concepts. The Schoolmen added that we use denominations present in the content of transcendentals in the analogical manner (the same idea is accepted by their contemporary followers). Unfortunately, I see no way for a satisfactory formal treatment of the analogy in this context. Perhaps, we should say that the meaning of transcendental concepts is not explicable through definitions, particularly established *per genus proximum et differentiam specificam*, but must be regarded as something primitive and intuitively graspable or perhaps captured by properly chosen axioms. Having no recommendable solution to this question, I would like to make a case study on being and truth, presumably understood as transcendentals in the classical sense.

The crucial issue is to explain the following formula:

(1) *Ens et verum convertuntur* (Truth and being are convertible).

According to (**) it means

(2) Being and truth are co-extensional.

Fortunately, we do not need to worry about definitions of the concepts involved in (2), i.e. being and truth. However, here I understand the latter (truth) as explained by the semantic theory of truth. More specifically, the locution '*A* is true', where *A* is a sentence (statement, proposition, etc.) is essentially elliptic and means:

(3) *A* is true in a model **M** under a specified interpretation **I**.

It is important to remember that although *A* is an element of an object-language **L**, (3) belongs to its metalanguage **ML**. The truth-definition based on the concept of satisfaction is not particularly relevant here, although we should remember its metalinguistic character. An obvious suggestion stemming from the foregoing explanations is that being is somehow related to the concept of model. I shall return to these assumptions below.

Paul Weingartner (see Weingartner 1979, Weingartner 2000) offered improved versions of traditional arguments for and against the convertibility of being and truth. The arguments against it are the following:

(A) (a) Things that are different and not mutually convertible;
    (b) But thought and being are different;
    (c) Thus, being and truth are not convertible, and (2) is false.
(B) (a) There are negative facts;
    (b) Negative facts are represented by true sentences;
    (c) True sentences representing negative facts, represent not being;
    (d) Thus, being and truth are not mutually convertible, and (2) is false.

On the other hand, we have the following argument for (2):

(C) (a) Every true sentence represents a fact (something that is the case) and vice versa;
(b) Truth is nothing other than which is the case;
(c) Hence, being and truth are convertible, and (2) is true.

Historically speaking, (A)(b) goes back to Aristotle (truth is not in being), (B)(b) was used by Brentano against the correspondence theory of truth, and (C)(b) has affinities with Augustine's view that *verum [...]esse id, quod est* (briefly, the truth is that what is).

Yet, there are some further problems concerning the interpretation of (A) – (C). Truth can be understood epistemologically or ontologically. For instance, the Stagirite opted for the former reading, while Augustine for the latter. The important point is that truth-bearers are ontological items, for example, things, according to the ontological approach, but if we adopt the epistemological theory, 'is true' is assigned to sentences, propositions, statements, thoughts, judgments, etc. Now, if truth is understood ontologically and is convertible with being, then, assuming that 'non-truth' refers to falsity, false being is not a being. In other words, 'false' is a modifier, not a determinator. Consequently, false truth is not a truth. This is forced by (*), but one can raise some doubts whether 'false' as applied to truth-bearers is always a modifier. In particular, on the epistemological reading, a true sentence (the same applies to other comparable truth-bearers) is a sentence, and the same concerns false sentences. Thus, 'false' is a determinator under the epistemological reading, and we have a clear difference between the epistemological and ontological understandings of truth. The problem which must be addressed in this context can be expressed by the following questions: Is the distinction between the epistemological and ontological accounts of truth relevant to (*)? And, *a fortiori*, are other versions of the convertibility thesis concerning being and truth relevant in this respect?

A solution that (2) applies to truth in its ontological understanding seems too simple if only for historical reasons. Although it is not clear how Thomas Aquinas and other Schoolmen understood truth, it seems that they argue for (*) as applicable to ontological as well as epistemological truth; they probably believed that *convertuntur* contributes to the explanation of such concepts as *adequatio, convenentia, conformitas* or *correspondentia*.[8] Thus, as far as the intensional relation between being and truth is concerned, theoreticians of transcendental maintained that if we said '*A* is true' and truth is defined as *adequatio rei et*

8 Fo a detailed presentation of truth-theories developed by Thomas Aquinas and and other Schoolmen see books by Aertsen quoted in footnote 1 and Schulz 1993.

intellectus (or by employing similar phrases), we added another parameter, i.e. a relationship between intellect and being. Thus, our truth-talk combines both ontological and epistemological features.

Since I will use the semantic concept of truth, the epistemological account is just my starting point, as Tarski adopted this perspective in his truth-theory. This is the first general assumption. Secondly, I repeat this constraint, truth is relativized to a model. Thirdly, the idea of correspondence does not play any role in my further considerations. Although it can be replaced by a precise semantic machinery, I will not use it here. Fourthly, models treated as algebraic structures substitute being. Yet, nothing particular is assumed about the ontological status of models. Of course, officially they are abstract mathematical objects. Fifthly, I assume that semantic models represent reality in the ordinary sense. In particular, the so-called standard model, in this case generated by the ordinary meaning of expressions, constitutes a faithful representation of the reality; we can consider it as the standard model of the real world. A simple but important consequence of this approach is that we do not need a special entity such as non-being or nothing as an ontological counterpart of 'being false', because sentences are true or false in models understood as ontological entities. Sixthly, the Gödel-Malcev completeness theorem (every consistent set of sentences has a model) guarantees that consistent sets of sentences have models, that is, ontological counterparts. Seventhly, there are three kinds of sentences: logical tautologies – true in all models, logical contradictions – false in all models, and logical contingencies – true in some models, false in other models. Now, single factual sentences are consistent and, thereby, have models – actual, if they represent reality or imagined, if they belong, for instance, to fiction. Eighthly, classical first-order logic serves as the logical base and it operates on denumerably infinite set of sentences.

Let **X** be a set of truths in a model **M**. It is a consistent set as it comprises only true sentences. By the Lindenbaum theorem, **X** has a maximal consistent extension, denoted by $\mathbf{X}^+$. So adding any sentence to $\mathbf{X}^+$ produces a contradiction. Hence, the collection (the term 'collection' is used intentionally) of all truths in **M** is unique. On the other hand, **X** as expressed in the first-order language has different models (the Löwenheim-Skolem theorem). Consequently, every model **M** has a unique (eventually up to isomorphism) collection of true sentences as its veridical (truthful or faithful) characterization. Furthermore, every set $\mathbf{X}^+$ has a collection $\mathbf{MOD(X^+)}$ – that is, the collection of all models of $\mathbf{X}^+$ – as its ontological (or semantic) counterpart. Furthermore, I will consider objects of the type $\mathbf{MOD(X^+)}$ in the above sense. We can formulate our problem in the following way: How are the collections of truths (true sentences) related to the collections of their model? Note that it is pointless to consider **X** and **MOD** (we **now** drop the upper index +) independently of each other. In fact, the Schoolmen had a fairly good intuition that truth adds something to being, namely the idea of a relation between them.

Set theory provides a simple argument for the transcendental character of being. Let us recall that being is represented by **MOD(X)**. The entity **MOD**(X) is a proper class, but not a set.[9] The reason is that the collection of all subsets of **MOD(X)** is larger than **MOD(X)**. If we consider sets as references of universals, **MOD(X)** is not an ontological counterpart of any universal concept. Clearly, **MOD(X)** cannot be defined *per genus proximum et differentiam specificam*, because proper classes cannot be subsets of other proper classes. This analysis nicely reproduces one of the most basic intuitions of the traditional theory of transcendentals. However, this argument is not automatically convertible to truth. In order to adopt it to the concept of truth, **X** should be considered as a proper class. There are arguments against doing that. Firstly, although the set *P*(**X**) of all truths in a given **M** is larger than **X** itself, it does not mean that it has more truths. It rather shows that truths from **X** can be grouped in more ways than the number of elements of **X**. Secondly, **X** is a subset of the collection of all sentences, true or false in **X**. This is a result of the fact that 'true' and 'false' are determinators under the epistemological understanding of truth. Alternatively, this larger collection is a proper class. If so, we have an argument for the transcendental character of 'true' or 'false', but not truth as such. The situation does not change if we assume that for any contingent sentence *A, A* is true in some model **M**, because there is still an over-collection of all sentences (truths + contradictions).

Assume that *A* and ¬*A* are both consistent. Hence, both have classes of models. Let us describe them as **MOD**{*A*} and **MOD'**{¬*A*} and assume that they are true in elements of these classes. We should not say that **MOD** or **MOD'** are true or false classes of models (or that they contain true or false models). The proper statement is that **M** (an element of **MOD**) is a model of *A* and that **M'** (an element of **MOD'**) is a model of ¬*A*. Or we can say – although this hardly adds anything new – that **M** is a true model with respect to *A*, a false model with

9 I use ZF (Zermelo-Fraenkel) set theory, because it provides the simplest solution concderning classes and their relations to sets. However, any other system in which the distinction of set and proper classes holds can be taken as a formal background. Fraenkel, Bar-Hillel, Levy 1973 and Arrigoni 1997, for technical accounts of set theories with classes.

respect to $\neg A$, while **M'** is a true model with respect to $\neg A$ and a false model with respect to A. This shows that 'true' and 'false', when relativized to models, are determinators, if the uses of both adjectives are coherent with the epistemological approach to truth. However, if a 'true model' is simply a model and 'false model' is not a model at all, 'true' becomes a trivializator, while 'false' a modifier.

It seems that there is no way to defend the convertibility thesis as a non-trivial thesis if truth is understood ontologically. In particular, 'ontological truth' is another label for being and we thus obtain that being is convertible to being and ontological truth is convertible to ontological truth. Epistemological truth itself is not convertible to being. If we take a specific true sentence, it corresponds to a model functioning as a being corresponding to it. However, the realm of being is not exhausted by sentences which are true in such and such models. On the other hand, there is a full convertibility of true or false (having a logical value) and being, because every sentence is true or false in some models. Hence:

(***) *ens et verum vel falsum convertuntur.*

Thus, the issue discussed here has a historical anticipation. It seems that the transcendentality of having a logical value was accepted by Frege and Wittgenstein, although they concentrated on truth. Frege was influenced by Neo-Kantian philosophers, and a Kantian reading of Wittgenstein's *Tractatus* is one of possible (and executed) interpretations of this book. However, this way of thinking essentially changes the tradition of transcendentality from scholastic to critical (Kantian).

Tarski's (and similar) construction of truth-definition causes that maxims like (***) must be expressed in a metalanguage. Perhaps, this circumstance deprives the traditional account of the convertibility of its required absoluteness. Due to the infinite hierarchy of languages and the infinite hierarchy of levels within the cumulative hierarchy of sets, the entire realm of truths and being cannot be defined. What remains is to point out that Truth as such and Being as such (the use of capitals is justified here) are ontological monsters or proper classes. However, because proper classes cannot be elements of something else, even other proper classes, a very limited number of predications can be applied to them. For example, if **C** and **C'** are proper classes, they are identical (equal, co-extensional) if and only if they have the same elements. Thus, assume that **C** is Being and **C'** is Truth. Consequently, **C** = **C'** if and only if every Being as such is true and every Truth is (exists). However, this formulation leads to all the problems discussed above. In particular, the Tarski undefinability theorem prevents us from defining the set of truths and consequently the corresponding class **MOD** even for a sufficiently rich object-language within it. Hence, in order to formulate (***) for a given language **L**, one must always draw on a suitable **ML**. The Tarski theorem can be considered as a modern version of the statement saying that transcendentals cannot be defined *per genus proximum et differential specificam.* On the other hand, for any level **L** (linguistic), we can define Truth and Being in **ML** and formulate a special case of (***). This is an optimistic conclusion, but it must be juxtaposed wwith a pessimistic one: from a universal perspective, our achievement is only partial. To quote Dante, *lasciate oggi speranza* that we can achieve something more than a very fragmentary ontological output, at least if simple logical facts are to be respected. This is a very significant lesson for formal ontology from formal semantics.

X
An Abstract Approach to Bivalence[10]

The principle of bivalence is usually expressed as the following statement ('either or' expresses the exclusive disjunction):

(1) every sentence is either true or false.

Any logic based on (1) is called two-valued or classical. Its traditional importance as the very foundation of formal logic consists in its relation to the so-called highest (the most fundamental) laws of thought. In classical metalogic, (1) is equivalent to the conjunction of two other basic rules of thought, namely the metalogical principle of excluded middle:

(2) every sentence is true or false;

and the metalogical principle of non-contradiction:

(3) no sentence is both true and false.

The adjectives 'true' and 'false' in (1), (2) and (3) are just metalogical (i.e. they belong to metalanguage and as such must be sharply distinguished from theorems $A \vee \neg A$, $\neg(A \wedge \neg A)$ and $(A \vee \neg A) \wedge \neg(A \wedge \neg A)$, that is, theorems of the propositional calculus formulated in the object-language of logic. Jan Łukasiewicz was perhaps the first logician who observed that (1) is metalogical (see Łukasiewicz 1930).[11] It is important to see that the equivalence

(4) (1) $\Leftrightarrow$ (2) $\wedge$ (3)

holds for classical logic only.[12]

A very general semantic approach to logic consists in adopting some logical values as designated and some as non-designated. Intuitively speaking, if a formula A always has a designated value as its logical value, it is a theorem of a given logic. Further, I will refer to sentences having designated values as designated sentences and to sentences having non-designated values as undesignated

---

10 This paper is a continuation of my earlier writings, namely Woleński 1998, Woleński 2004. In fact, my approach is not the most abstract as compared with account offered in Shramko and Wansing 2011.

11 See Betti 2001 for a discussion of Łukasiewicz's views on (1).

12 See Béziau 2004 for a discussion of relations of various logics to (1), (2) and (3). I will discuss his suggestions in a special digression a the end of this paper.

sentences. Assume that a function *val* (a valuation function) from **L** being a set of sentences to a set **V** of logical values is defined. There are three cases ('card' is the abbreviation for 'cardinality'):

(5) card(**V**) = **0**; no *A* is a theorem;
(6) card(**V**) = **1**; all *A's* are theorems (a given logic is inconsistent);
(7) card(**V**) ≥ **2**; some *A's* are theorems, some *A's* are not theorems.

Let **D** be a set of designated values. A logic **Lg** is consistent if card(**V**) > card(**D**) (a necessary condition). Moreover, if *A* is not a theorem, ¬*A* is contradictory (absolutely or relatively, depending whether the issue concerns pure logic or extralogical theories). In the case of a complete logic, *A* is contradictory iff *A* is always non-designated.

Clearly, (1) displays the minimal situation where we have exactly two values. It is normally assumed that truth is designated and falsity is non-designated. On these grounds, *A* is a tautology iff *A* is true for all valuations, and a counter-tautology iff *A* is counted as false by all valuations. At the same time, however, dual logic (and other similar constructions) designates falsehood. This shows that designating truth is based on some pragmatic assumptions even in the most popular case of two-valued logic with truth as the designated value. Anyway, the sets of designated and non-designated values are mutually exclusive in any valuation. Moreover, both sets exhaust **V** (for simplicity's sake, I consider truth-value gaps as logical values in this context). And the division of logical values into designated and non-designated forms the logical division in the traditional understanding (see Greniewski 1970a; I skip the difference between the bivalentists and the pseudo-bivalentists introduced by this author). The problem of stability of having definite logical values (the question whether sentences change their values, for instance, in the course of time or dependently on the region of space in which facts described by the sentences in question occur) is more complex and I will omit it here.

On the other hand, there is no a priori reason to restrict **V** to two items. Consider the logical square (**D1**) with truth as modality (**T** – is true):

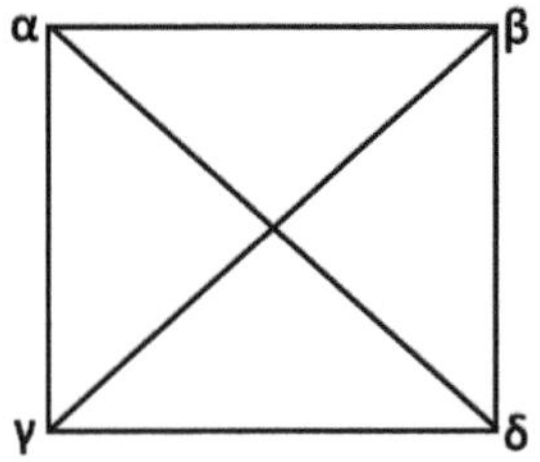

where particular points have the following interpretation: α – **T***A*, β – **T**(¬*A*), γ – ¬**T**(¬*A*), ¬**T**(*A*). We have the following dependencies (analogical to the relations between modalities):

(8)	$\alpha \Rightarrow \gamma$	(truth of *A* implies truth of non-truth of ¬*A* is true);
(9)	$\beta \Rightarrow \delta$	(truth of ¬*A* implies non-truth of *A*);
(10)	$\neg(\alpha \wedge \beta)$	(truth of *A* and truth of ¬*A* are contraries);
(11)	$\gamma \vee \delta$	(non-truth of ¬*A* and non-truth of *A* are complementarities);
(12)	$\alpha \Leftrightarrow \neg\delta$	(truth of *A* and non-truth of *A* are contradictories);
(13)	$\beta \Leftrightarrow \neg\gamma$	(truth of ¬*A* and non-truth of ¬*A* are contradictories).

There is, however, a question where falsehood should be located in the diagram (**S1**) (the formula **F***A* abbreviates "it is false that *A*" or "*A* is false"; I regard the operator and the predicative formulation as expressing the same proposition). There are two possibilities, namely to place **F***A* either at the point β or at the point δ. The first choice equates **F***A* with **T**(¬*A*), while the second decision identifies **F*A*** with ¬**T***A*. In fact, (1) assumes that **F***A*, **T**¬*A* and ¬**T***A* are equivalent. Putting it formally, we have the following equivalences

(14) (a) T¬A ⇔ ¬TA;
(b) FA ⇔ T¬A (or ¬TA (or ¬TA).

This is a tacit assumption of the traditional principle of bivalence.

However, (**D1**), unless (14) is accepted , does not preclude that thera are *A's* which are neither true not false. It is illustrated by means of the logical hexagon (**D3**):

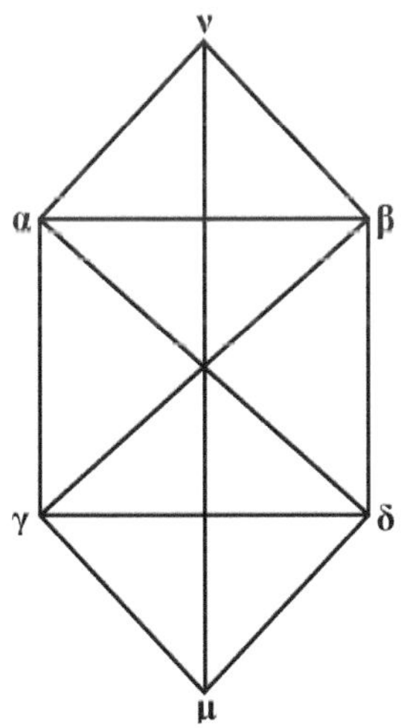

Additional points ν and μ have the reading as respectively "*A* is either true or false" and "*A* is neither true not false" provided that falsehood is located at the point β; this choice seems the most intuitive from the general point of view. If (14) is accepted, (**S2**) becomes reducible to the triangle determined by the points αβν and we can universally generalize ν. This gives

(15) $\forall A(\mathbf{T}(A) \vee \mathbf{F}(A))$,

that is (1). Clearly, the principle of bivalence in this construction is not a necessary (analytic, metalogically tautological) statement, because it must be justified by stipulations recorded in (14); the mentioned reduction fails, if (14) is rejected.. This observation immediately supports Łukasiewicz's view that the principle of bivalence goes beyond purely logical reasons. And what about the point μ, if (14) is rejected? There are several interpretative possibilities, for instance, (a) to admit the existence of truth-value gaps (sentences devoid of logical values; some logicians point out that the Liar sentence is gappy); (b) to allow other logical values than **T** and **F**, or (c) to introduce fuzziness. Such solutions lead to various reforms of classical logic, for instance, by introducing many-valued logic, logic with truth-value gaps or systems formalizing fuzzy (inexact, rough) concepts. The details of such constructions are fairly complicated. For example, fuzzy logic is frequently considered as a special kind of many-valued logic. We can also regard the gappiness as a separate logical value (this pragmatic solution was adopted above for simplicity). My further considerations are restricted to (b), except for some illustrations. It is sufficient to consider the three-valued system with **T** as the designated value.

It is possible to interpret (**S1**) and (**S2**) by replacing **T** by **D**. In this context, **D** refers to a designated value, not to the set of designated values. Thus, the formula **D***A* means "it is designated that *A*", "*A* is designate" or "*A* is a designated sentence". We can formulate the principle of bivalence as:

(16) $\forall A(\mathbf{D}(A) \vee \neg \mathbf{D}(A))$,

which expresses the proposition that every sentence is designated or non-designated (see Kotarbiński 1913). For instance, every sentence is true or (false or neutral), every sentence is true or (false or gappy). (16) seems trivial, although it is universally valid, according to the logic of generalized logical squares (see Woleński 2008 for an examination of such diagrams). In particular, the category of non-designated items is ambiguous or heterogenic, because it covers falsehoods and neutralities (or gappy sentences). In fact, this solution is too easy, because if we have a given non-void universe **U** consisting of three (or more) mutually exclusive sets *X*, *Y*, *Z*, that is, $\mathbf{U} = X \cup Y \cup Z$, it is possible to form the set $X' = Y \cup Z$ and transform the initial equality into $\mathbf{U} = X \cup X'$; the symbol X' refers to the complement of X.

In order to refine the picture, we can divide non-designated sentences into anti-designated (Rescher 1969, Malinowski 1993, Gottwald 2001, Shramko, Wansing 2011), $\mathbf{D}^{\mathbf{A}}$ and simply non-designated $\mathbf{D}^{\mathbf{S}}$. The rule is that if $\mathbf{D}^{\mathbf{A}}(A)$, then $\mathbf{D}^{\mathbf{S}}(A)$, but not conversely. Moreover, $\mathbf{D}(A)$ and $\mathbf{D}^{\mathbf{A}}(A)$ are contraries, but $\mathbf{D}(A)$ and $\mathbf{D}^{\mathbf{S}}(A)$ – contradictories. Simply non-designated sentences can have designated values in some models and non-designated values in other models. Having these dependencies, we can formulate a non-trivial generalized version of the principle of bivalence as:

(17) $\forall A(\mathbf{D}(A) \vee \mathbf{D}(\neg A))$,

which says that every sentence is either designated or anti-designated. Clearly, (15) – and, *a fortiori*, (1) – becomes a special case of (17), if T and F are admitted as the only logical values and truth is designated, but falsehood functions as anti-designated. Moreover, there is no difference between being anti-designated and non-designated for the latter. On the other hand, (17) also holds if F serves as the designated value (dual logic). Clearly, the principle (17) is not-tautological, because it is false in some metalogics.

If T is designated, its behavior is governed by the T-scheme:

(18) $\mathrm{T}A \Leftrightarrow A$,

but we cannot replace T by $\mathbf{D}$ in (18). To see it, consider the diagram (**D3**).

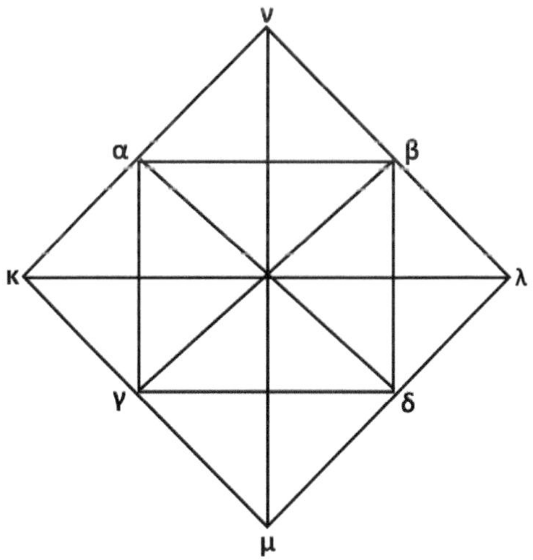

The points κ and λ refer to A and $\neg A$, respectively. Although,

(18) $\mathbf{D}A \Rightarrow A$,

holds for any designated value, because if it is true that A is designated, A represents its designated value and its conversion generally fails. In fact, if a sentence A represents a value v, it does not mean that it is designated. For instance, if *val* ascribes the neutrality value n to A in the three-valued system of Łukasiewicz with **T** as designated, we still have that $\mathbf{D}A \Leftrightarrow A$, but $n(A)$ does not imply $\mathbf{D}A$.

The above analysis suggests that Tarski had a good intuition when he characterized the concept of truth by applying **T**-equivalences. The propositional calculus is the simplest case of logic. We can introduce truth and falsehood as unary connectives by equations (**1**, **0** refer to values of the function *val*)

(19) $v(\mathrm{T}A) = \mathbf{1}$ iff $v(A) = \mathbf{1}$; otherwise $v(A) = \mathbf{0}$.

These equations and earlier observations – however simple or perhaps even trivial they might seem – are interesting for at least four reasons. Firstly, they show why (1) is frequently confused with the law of excluded middle. In fact, (19) justifies the reduction of $\mathrm{T}A \vee \neg \mathrm{T}A$ (or $\mathrm{T}A \vee \mathrm{T}\neg A$ or $\mathrm{T}A \vee \mathrm{F}A$) to $A \vee \neg A$. Yet, this is only a partial success, because if we move to the predicate calculus, it turns out that introducing "it is true" as a monadic connective does not suffice as the foundation of semantics and the full-blooded truth-definition based on the concept of satisfaction must be employed. Secondly, (19) additionally explains why (**S2**) and (**S3**) cannot be reduced to the triangle αβν in the case of admitting neutralities or truth-value gaps. Thirdly, considerations about the T-scheme show that T-equivalences are no longer logical tautologies beyond the propositional calculus (see Woleński 2008c for further arguments). Fourthly, (16), 17) and (19) contribute to an analysis of the Suszko Thesis (see Suszko 1977),[13] according to which every logic is bivalent. More specifically, the Suszko Thesis distinguishes logical values (T, F) and algebraic values (for instance, neutralities in three-valued logic). The simplest argument saying that this distinction is conventional consists in pointing out that every logic can be made bivalent by (16), but not by (17). In other words, and on purely logical grounds, the Suszko Thesis is problematic even if we replace T by **D**. In fact, it is trivial for **D** and $\mathbf{D}^{\mathbf{S}}$, but not for **D** and $\mathbf{D}^{\mathbf{A}}$, although some abstract algebraic constructions somehow "de-trivialize" the problem. Nonetheless, it is dubious whether they remain inside logic.

13 See also Woleński 2009a, Caleiro, Carnielli, Coniglio, Marcos 2007; while I am skeptical about Suszko's thesis, the Brazilian authors defend it.

The formula (17) holds for any two-valued case (the paraconsistent case will be considered separately below), for instance, when F serves as the designated value. Yet, logical investigations much favour (15), particularly in metalogic. Moreover, metalogical studies are based on classical logic even if they pertain to non-classical systems. Although there are various attempts to execute non-classical metalogic, the results are fragmentary and limited in their applications. For instance, important metatheorems concerning intuitionistic logic have no proofs by which would use methods accepted by the intuitionsts; the completeness theorem for the intuitionistic predicate calculus is a very spectacular example in this respect. In fact, no general metatheory of logic based on non-classical logic is available. Although this conclusion has the status of an empirical generalization, it suggests that non-classical logics arise when some refinements of classical logic help to shed some light on specific problems (constructive proofs, fuzziness, gaps, future contingents, paradoxes, quantum mechanics, etc.). Anyway, the truth-talk has a privileged status in metalogic and there is nothing indicating that the situation will change. For example, logicians speak about truth-functions although the label "falsehood-functions" appears as fully legitimate from the theoretical point of view.

What are the reasons for seeing bivalence as basic and truth as designated? Apart from the pragmatic motives pointed out in the last paragraph, we can list the following circumstances (I restrict myself to a stock of some factors without entering into details; the list below is nothing more than a body of speculative suggestions):

(a) Truth represents facts, but falsehood indicates the lack of facts. People are more interested in facts than in the absence of facts (this point was made by Jerzy Perzanowski in one of our discussions about the concept of truth);
(b) Spatial and temporal oppositions (symmetric or not): left–right, bottom–top, earlier–later, back–front, outer–inner;
(c) Biological oppositions: life–death;
(d) Motorical oppositions: active–passive;
(e) Other distinctions: having something – having nothing, having all (in some collection) – having not-all; modal contrasts and dualities, differences in skills, utilities, duties, values, attitudes, etc., for instance, duty–right or useful–non-useful. Observe that some oppositions involve positive and negative contrasts, while other are based on the distinction between the positive and the privative;
(f) Binary rhythms;
(g) Sensual contrasts: dark–bright, loud–silent, colourful–colourless;
(h) The structure of double helix;
(i) Binary character of genetic codes;
(j) 1–0 nature of information;

(k) The belief that preserving truth help protecting information, but preserving falsehood is a "measure" of information-dispersion;
(l) The dual behaviour of many quantifiers in the natural language.

For these and presumably many other reasons, logic acting as the framework of human thinking in a notoriously contrastive environment seems to require a minimal world as its ontological counterpart and this is the world with two items, i.e. logical values. For pragmatic reasons, one component was called Truth and the other was dubbed Falsehood. Let me conclude with an anecdote. Several years ago, perhaps at the beginning of the 1970s, Roman Suszko delivered a talk on the Non-Fregean logic. He argued that the Frege axiom asserting that all true propositions refer to one object, namely the True, and all false propositions denote the False as their very reference, is not intuitive and should be rejected. According to Suszko, propositions refer to situations and we know that there are many situations. In the discussion held after the talk, Andrzej Wroński remarked that the logical world should be as limited as possible, that is, to two items. Clearly, (7) shows the minimal furniture of the logical reality. Although the principle of bivalence with **T** as the designated value bears no stamp of logical necessity, we can give several empirical reasons to support it.

A Diggression

Jean-Yves Béziau (see Béziau 2004) derives the principle of bivalence from the following statements (I use my terminology):

(i) The set of logical values is limited to two values;
(ii) These truth-values are truth and falsehood;
(iii) *Val* is a function;
(iv) *Val* is a total function.

The points (i) – (iv) are summarized in one statement (Béziau 2004, p. 74):

> The evaluation relation is a total function whose domain is the set of propositions and the codomain is a set of truth-values, true and false.

Clearly, (i) excludes many-valueness, (ii) identifies truth-values, (iii) states the standard property of *val* and (iv) excludes truth-value gaps.

Béziau's main intention is to show that (1) should not be identified with the conjunction of (2) and (3), because, on the one hand, we have a bivalent semantics for logic without the principle of excluded middle, and, on the other hand, there is a bivalent semantics for logic without the principle of non-contradiction. Béziau's makes several observations, for instance, he says (p. 76) that bivalence "reappears at another level" in many-valued logic, namely as the principle expressed by (16). However, as I pointed out, this is a quite trivial version of (1) as compared with (17). Using Béziau's terminology, (16) obeys the principle of bivaluation, but not the principle of bivalence. Thus, I exclude bivaluational semantic constructions as necessarily bivalent (in my sense).[14] Moreover, Béziau toys with the principle of excluded middle (EM, for brevity). First of all, it is not always sufficiently clear whether he appeals to its logical or metalogical version, as both versions appear in his considerations. Furthermore, Béziau offers a non-standard formulation of the principle in question, namely

(EM) A sentence A and its negation cannot be both false.

Béziau argues that (1) does not entail (EM), because there is a bivaluation ascribing **0** to A as well as to $\neg A$. An answer to this argument points out that (i) – (iv) do not impose truth-conditions for connectives. Yet, the classical bivalent semantics as contrasted with various bivaluations defines the functor of negation in the manner used in (1).

The paraconsistent case is perhaps particularly interesting. Doubtlessly, paraconsistent semantics obeys (i) – (iv). The third condition allows that both A and $\neg A$ can have the value **1** in some paraconsistent systems or the value **0** in other systems. The first case concerns so-called dialetheias (contradictory statements which are both true), but the second case tolerates anti-dialetheias (the name is *ad hoc*), that is, ascribes falsehood to both mutual contradictories.[15] Moreover, we can designate **T** (or **F**, if one likes), define a non-explosive consequence operation and determine a sound "bivalent" logic (the quotes here will be explained in a moment). However, there are some survey problems to be tackled. First of all, dialetheias (and anti-dialetheias as well) cannot be generally generated, but they arise only from very special cases. These are, for instance, the Russell sentence about sets which are not their own elements, the Liar sentences, etc. One may ask what is the difference between such sentences and "normal" contradictories. Secondly, I did not find any natural matrix semantics for paraconsistent logic.[16] Hence, properties of connectives are somehow vague or even mysterious. For instance, the negation in discussive logic is classical, but the negation in one of Béziau's examples is not. Thirdly, the relation of paraconsistent logic to many-valued logic is not clear. Some logicians (see Priest

14 See Caleiro, Carnielli, Coniglio, Marcos 2007 for a very sophisticated account of dyadic (bivaluational) semantics. However, constructions presented in this paper seem to be based on the distinction of designted and non-designated values.

15 Dialetheieas and anti-dialetheias are someitems called truth-value gluts.

16 Compare for instance Béziau, Carnielli, Gabbay 2007 or Priest, Tanaka 2009.

and Tanaka 2009 for further information) regard many-valueness as the best way to explain paraconsistency, but other tend to search for different solutions. Fourthly and more specifically, it is very unclear how the **T**-scheme should be understood if we insert dialetheias (or anti-dialetheias) in it. In the former case, it is very problematic whether paraconsistency obeys the principles of bivalent semantics, but if the latter direction is taken, the issue appears to be open. These arguments justify the hypothesis that paraconsistency is bivalent if and only if it is bivaluational. In fact, although bivalence leads to a bivaluation, the reverse dependency does not generally hold. Let me add that my critical remarks about paraconsistent logic are not intended as offering arguments against formal studies on consistency, its sources, scope and limitations. All I wanted was to show why the standard concept of bivalence should not be extended beyond classical logic.

XI
Philosophical Reflections on Logic, Proof and Truth

Let us agree that logic (for simplicity, I limit my further considerations to classical first-order logic and theories; thus, non-classical, deviant or alternative logics, including infinitary systems, and respective theories based on such logical systems are entirely ignored) codifies the methodology of mathematics. For Frege, logical laws are inherently connected with truth. He says (Frege 1979a, p. 128 and Frege 1979b, p. 3):

> How must I think in order to reach the goal, truth? We expect logic to give us the answer to this question, but we do not demand that it should go into what is peculiar to each branch of knowledge and its subject-matter. On the contrary, the task we assign logic is only that of saying what holds with the utmost generality, whatever its subject-matter. We must assume that the rules for our thinking and for our holding something to be true are prescribed by the laws of truth. The former are given along with the latter. Consequently we can also say: logic is the science of the most general laws of truth.
> Logic is concerned only with those grounds of judgment which are truths. To make a judgment because we are cognizant of other truths as providing a justification for it is known as *inferring*. There are laws governing this kind of justification, and to set up these laws of valid inference is the goal of logic.

However, Frege's view appears as not quite coherent. To perform a valid inference from A to B, means to deduce B from A, that is, to demonstrate that A logically entails B. Otherwise speaking, it means to produce a (deductive) proof of B on the basis of A as the assumption. Such an inference cannot lead from A as true to B as false. All other possibilities (A – true, B – true; A – false, B – true; A – false, B – false) are admitted and do not preclude the validity of concrete proofs. Hence, the proof (I equate 'proof' and 'valid proof') of B from A (symbolically, $A \vdash B$ or, more generally $X \vdash B$, where X is a set of premises) does not depend on the truth of A. If we want to add the claim that premises of a proof should be true, we should make a further distinction, namely to distinguish materially correct proofs, that is, inferences having true premises and formally correct proofs, that is, deductions which are not dependent on this constraint and defined as respecting logical inference rules. In fact, Frege's seminal logical works codify formally correct proofs, not materially sound ones.

Hilbert's formalism (Hilber, Ackermann 1928, p. 1; reference to Eng. tr.] considers logic as interested in

The transition from statements to their consequences, as occurs in the drawing of conclusions, is analysed into its primitive elements, and appears as a formal transformation of the initial formulas in accordance with certain rules, similar to the rules of algebra; logical thinking is reflected in logical calculus.

This explanation does not apply to the concept of truth. Thus, logic, on Hilbert's view, is exclusively interested in the laws governing the concrete instances of the relation $\vdash$. Most explanations of what is logic follow the last quotation. Here is a sample taken from elementary as well as advanced sources:

(a) Symbolic or formal logic [...] is the study of the various general types of deduction.
(b) [...] [formal] logic is concerned with the analysis of sentences or of propositions [...] and of proofs [...] with attention to the *form* in abstraction from the *matter*.
(c) The aim of constructing our symbolic logic is that is shall serve as a precise criterion for determining whether or not a given instance of [...] reasoning is correct.
(d) The primary subject-matter of logic is the structure pattern of demonstrative inference.
(e) [...] logic is [...] the analysis of methods of reasoning [...] logic is interested in the form rather than the content of argument.
(f) Symbolic logic is a mathematical model of deductive thought.
(g) Logic is the study of correct reasoning.
(h) Formal logic is [...] a theory of logical consequence.[17]

All (a) – (h) are fairly similar and explain that logic deals with deductive, correct, valid or demonstrative inferences and does exclusively that by reference to formal properties of arguments. Obviously, the terms 'valid argument', 'correct reasoning", 'deductive reasoning' or 'inference' function as synonyms in (a) – (h) or, at least, as co-extensive labels.

How truth appears in logic under (a) – (c) and similar elucidations? One possibility is that we will speak about so-called logical truths, that is, universally valid formulas. Frege's view seems to license this possibility, because he considered theorems of logic as analytic sentences, that is, sentences provable by rules of logic and definitions. However, this solution appears as too restrictive or even trivial. Formal (logical) truth and material truth were traditionally contrasted. Frege's account of the goal of logic certainly concerns material truth. Now, since the passing from logical theorems (i.e. provable formulas) to logical truth (i.e. universally valid formulas) uses the weak completeness theorem (see below), it assumes the concept of material truth. Otherwise speaking, if we define logical truth as true in all models (or true under all interpretations), we tacitly appeal

17 (a) Russell 1903, p. 10; (b) Church 1956, p. 1; (c) Rosser 1953, p. 5; (d) Kneebone 1963, p. 5; (e) Mendelson 1964, p. 1; (f) Enderton 1972, p. 1; (g) Bonevac 1990, p. 2; (h) Pogorzelski 1994, p. 19.

to sentences which are materially true. Another contrast (see Van Heijenoort 1967), going back to Leibniz, namely of logic as *characteristica universalis* and logic as *calculus ratiocinator*, is to some extent helpful in the present context. This distinction was recently refreshed (see Hintikka 1988, Woleński 2012) as concerning logic as calculus and logic as language. Although the former sees logic as the battery of rules of inference, it is not quite clear how can the concept of truth be incorporated into *lingua characteristica* (see Kusch 1989, Woleński 2004a for further remarks).

Frege worked within the tradition of *lingua characteristica*. His view appeals to the concept of material truth and he considered it as non-definable. On other hand, Hilbert explicitly considered logic a calculus (see the quotation above), but the concept of satisfaction played a crucial role in his metamathematical investigations. Thus, if we place the concept of truth in semantics, who was a semanticist: Frege or Hilbert? Book titles very frequently reflect the needs of marketing. But perhaps it is not accidental that we can read the following books: *An Introduction to Mathematical Logic and Type Theory: to Truth through Proof* (Andrews 1986) or *What Logics Mean. Truth through Proof. A Formalist Foundation of Mathematics* (Weir 2010) or *From Proof Theory to Model-Theoretic Semantics* (Garson 2013). These books, written from the perspective of different philosophies of mathematics, do not follow Frege and his idea that the goal of logic consists in reaching truth. What one could derive from the content of the mentioned volumes, is that proof theory without model-theoretic semantics is somehow too poor to display the essence of logic (at least I arrive at this conclusion). Perhaps, the succession that we proceed from proof to truth has a significance. I will try to demonstrate that it has. Historically speaking, Hilbert's formalism, (at least as performed in Hilbert, Bernays 1934-1939; see also Herbrnad's papers collected in Herbrand 1971) recommends proceeding from intuitive (material truth) via formalization and formal finitary proofs to truth. On this account, finitary proofs act as *the* criterion of truth via consistency. Gödel started as a formalist, but his famous results on incompleteness suggested that the concept of truth cannot be entirely eliminated. Tarski partially followed this path in the 1930s, but he formally defined the concept of truth and showed the reasons explaining why there is an essential and deep gap between truth and proof (provability).

In order to compare truth and proof, we need locate both concepts on the same level. Logic as codified by Frege and Russell was mostly expressed in the object-language, although the rules of inference implicitly refer to provability. Since Hilbert introduced metavariables and schemata instead concrete object-language formulas, his formalization became more metatheoretical., but it still can considered as a certain calligraphic variant of the traditional way of presenting logic. Systems of natural deduction (Jaśkowski, Gentzen) and the sequent calculus (Gentzen) were more metalinguistic, because they proceeded not by producing theorems, but by modes of derivations. Of course, passing from proof-figures to theorems is always possible, for instance, via the deduction theorem. One might say that natural deduction and the sequent calculus merge object language and metalanguage to some extent. However, if we extract the provability relation, only a few of its properties can be defined inside logic itself; for example, the cut-rule generates one of such properties. As Tarski showed, the concept of truth and other semantic notions are essentially metatheoretical. Contemporary metamathematics covers proof theory (metamathematics in the Hilbertian sense) and model-theory (semantics in the Tarskian sense). Hence, a comparison of truth and proof can be effectively realized at the level of metamathematics, not pure logic as expressed in the object-language or even in its metalignguistic limitation. This is the crucial point, which also allows us to extend the issue from logic to arbitrary theories based on logical machinery as the inference instrumentarium.

A proof of a formula A from the premises $\{A_1, A_2, ..., A_n\}$ is a finite sequence of formulas $A_1, A_2, ..., A_n, A_{n+1}$ such that $A_{n+1} = A$ and every step from A_i to A_{i+1} ($1 \leq i \leq n$) has a justification in admissible rules of inference (in particular, every axiom constitutes its own proof; I omit further details present in the full definition of proof). We write (see above) $A_1, A_2, ..., A_n \vdash A$ if A is provable from the assumptions $A_1, A_2, ..., A_n$. Another notation is based on the definition of the consequence operation Cn:

(*) $A_1, A_2, ..., A_n \vdash A$ iff $A \in Cn\{A_1, A_2, ..., A_n\}$.

If $A_1, A_2, ..., A_n \in X$, we can also write $A \in CnX$. The general axioms for Cn as established by Tarski are as follows:

(C1) the language **L** on which Cn operates, is at most infinitely denumerable and consists of formulas of the finite length);
(C2) $X \subseteq CnX$ (every set is a subset of its consequences);
(C3) $X \subseteq Y \Rightarrow CnX \subseteq CnY$ (Cn is montotonic);
(C4) $CnCnX = CnX$ (Cn is idempotent);
(C5) $A \in CnX \Rightarrow \exists Y \subseteq X \wedge Y \in \mathbf{FIN} \wedge (A \in CnY)$ (Cn is finite);
(C6) $(A \Rightarrow B) \in CnX \Rightarrow B \in Cn(X \cup \{A\})$ (the deduction theorem).

The axiom (C6) leads to the definition of logic as the set of consequences of the empty set (**LOG** = $Cn\varnothing$). This definition, perhaps slightly artificial, is equivalent (see Surma 1981) to other plausible statements, for instance, $A \in$ **LOG** iff $\neg A$ is inconsistent or **LOG** is the only non-empty product of all deductive systems. (C5) establishes that if $A \in CnX$ there exists a finite subset Y of X such that A is a consequence of Y. This property is also expressed by the definition of the operator $\vdash$ because the sequence $\{A_1, A_2, ..., A_n\}$ is always finite. In fact, we can define consequence operations for infinite sets of premises or for languages with infinitely long formulas (so-called infinitary languages) but such proofs, although they can be abstractly described, transcend human inference capacities. Otherwise speaking, only finite proofs in the sense of (C1) and (C5) are checkable. Yet, the general possibility of checking in the finite number of steps does not guarantee that proofs are finitary in Hilbert's sense, constructive in the intuitionistic sense or even practically realizable.

A theory **T** is a set of formulas (sentences, that is, formulas without free variables) such that **T** is closed by *Cn*. Symbolically, $\mathbf{T} = Cn\mathbf{T}$; since the inclusion $\mathbf{T} \subseteq Cn\mathbf{T}$ is given in (C2), the dependence $\mathbf{T} \subseteq Cn\mathbf{T}$ displays the main property of theories in the logical sense (**T** is a theory if it contains its own consequences). Logic is one of theories. Since $\varnothing \subseteq \mathbf{T}$, (C3) immediately gives that logic belongs to consequences of any theory, and the definition of **T** entails that logic as a proper part of any extralogical theory. Hence, we can say that *Cn* codes the inferential machinery of theories associated with *Cn* (recall that I assume the classical logic; it can be defined by adding specific axioms to (C1) – (C6) but I omit this supplement).

The weak completeness theorem says: $A \in \mathbf{LOG}$ ($= Cn\varnothing$) iff A is true in every model **M**, and the strong completeness theorem is displayed by the formula $A \in \mathbf{T}$ ($= Cn\mathbf{T}$) iff A is true in all models of **T**. The implication (I omit the weak completeness theorem) '$A \in \mathbf{T} \Rightarrow A$ is true in all models of **T**' and expresses the soundness of logic (the deduction does not produce falsehoods from truths), but the implication 'A is true in all models of $\mathbf{T} \Rightarrow A \in \mathbf{T}$' establishes the proper completeness theorem and says that every **T**-truth is provable. An immediate suggestion from the proper completeness theorem is that proof constitutes a criterion of truth. Thus, the slogan "from proof to truth" can be considered as a general lesson from the completeness theorem. Of course, the Hilbertian formalist adds "yes, but proofs should be finitary". The completeness theorem of the above version is equivalent to the following assertion (the Gödel-Malcev completeness theorem): a theory **T** (more generally, an arbitrary set of sentences) is consistent iff it has a model. This nice equivalence would satisfy the formalist provided that all ingredients of it could be provable by finitary devices. To complete the preliminaries, I understood truth as truth in a model **M**, that is, according to Tarski's semantic definition of this concept (technical details do not play any essential role here).

We have two objects $\mathbf{T}^{\mathbf{Cn(T)}}$ and $\mathbf{T}^{\models(\mathbf{M})}$. The former refers to theories in the syntactic sense (defined by the concept of the consequence or provability), but the latter, which is semantic (the sign $\models$ (**M**) symbolizes truth in **M**), sees theories as true (or satisfied) in a class of models). The strong completeness theorem establishes the parity of both concepts. Moreover, the Gödel-Malcev theorem equates consistency and having a model in that it says that if a theory **T** is consistent, it has a model in which its theorems are true. Unfortunately (for the formalist), these assertions have serious limitations. Firstly, the strong completeness theorems as well as its equivalent (the Gödel-Malcev theorem) have no constructive proofs (yet, their equivalence can be effectively established). This means, among other things, that in some cases, although we know that a theory **T** has a model, we have no (or an insufficient) idea how to build it. More strictly speaking (this conclusion involves additional assumptions), syntactical devices are effective (in the sense of being recursively enumerable), but semantics is not. And using a philosophical jargon, semantics transcends syntax as far the issue concerns the effective character of the latter, but the non-effective nature of the former. Thus, the realization of the slogan "from proof to truth" encounters serious difficulties under formalistic constraints as far as the completeness theorem is involved.

The concept of completeness as defined above must be distinguished from the syntactic concept of completeness: **T** is syntactically complete iff for any A (expressible in **L** of a given theory **T**), $A \in Cn\mathbf{T}$ or $\neg A \in \mathbf{T}$ (or more simply, $A \in \mathbf{T}$ or $\neg A \in \mathbf{T}$). Some theories, for instance, the Presburger arithmetic (arithmetic with addition as the sole operation) or the Skolem arithmetic (with multiplication as the sole operation) are syntactically complete, but most interesting mathematical (and, in particular, empirical theories based on arithmetic) are lacking of the syntactic completeness. This important fact was proved by Gödel in his first incompleteness theorem. The Peano arithmetic (arithmetic of natural numbers with addition and multiplication denoted by the symbol **PA**) is demonstrably incomplete, that is, there exists an arithmetical sentence A such that $A \notin \mathbf{PA}$ (= $Cn\mathbf{PA}$) and $\neg A \notin \mathbf{PA}$ (= $Cn\mathbf{PA}$, provided that **PA** is consistent. Supplementing **PA** by adding to it an unprovable (undecidable) sentence A as a new axiom, does not improve the situation, because the resulting theory $\mathbf{PA} \cup \{A\}$ remains incomplete as well. Thus, **PA** and its extensions inherit inevitable incompleteness from the former (I will speak about **PA** and its properties, but it automatically applies to all extensions of this theory). Furthermore, the sentence asserting that **PA** is consistent (con(**PA**)) cannot be proved in **PA** (Gödel's second incompleteness theorem saying that con(**PA**) $\notin$ **PA** and $\neg$con(**PA**) $\notin$ **PA**).

The formula con(**PA**) is eventually provable in richer systems (this game might be continued *ad infinitum*), but an appeal to infinistic methods cannot be eliminated. Hence, since arithmetic seems to produce the upper limit of what is fully reliable (that is, diminishing the risk of inconsistency), the situation is actually dramatic: consistency of arithmetic cannot be proved by reliable proof methods. In fact, the issue is still controversial for the lack of a commonly accepted definition of mathematical reliability. Some authors do not admit all arithmetical techniques as reliable, but other permit some infinistic methods. I adopt a kind of compromise, but the essential point is that the problems will reappear on every stage higher than **PA**, but, on the other hand, hopes to prove con(**PA**) in a weaker theory appear as demonstratively unrealistic. Yet, this does mean that non-constructive consistency proofs have no importance. Firstly, they are mathematically interesting and, secondly, they can serve as empirical data justifying faith in consistency. Anyway, the incompleteness phenomena displayed by both Gödel's theorems essentially enrich the lessons coming from the completeness results. We know that the strong completeness theorem (and, *a fortiori*, the Gödel-Malcev theorem) do not admit constructive proofs. Clearly, if **PA** has a model, it is consistent, but models (there are many models; remember that I take into account first-order theories) of **PA** cannot be built by constructive devices. Accordingly, the fate of the formula 'con(**PA**)' explains the reasons why the strong semantics completeness theorem cannot be proved constructively for first-order theories.

The first Gödel theorem has also other formulations, namely (a) The set of arithmetical truths is not finitely axiomatizable; (b) There are true but not provable arithmetical sentences. Both these formulations are particularly interesting, because they use the term 'true'. Gödel thought about truth in an intuitive way. In particular, he did not define the concept of truth and relied on its intuitive understanding. On the other hand, the method of arithmetization envisaged by Gödel for proving both incompleteness theorems (in fact, the second is a consequence of the first) leads to a definition of the concept of proof within **PA**, that is, via effective devices. Consequently, we can define the set of provable numbers (this term was suggested by Tarski; see Tarski 1969; this paper considers proof as the criterion of mathematical truth), that is, Gödel numbers of provable sentences. Gödel was aware that the concept of truth behaves in another way (Gödel 1934, p. 363):

> So we see that the class α of numbers of true formulas cannot be expressed by a propositional formula of our system, whereas the class of β of provable formulas can.

This statement can be interpreted as a confession that the concept of truth transcends **PA**. The issue was completed by Tarski and his undefinability theorem (see Woleński 2004 for an extensive comparison between Gödel and Tarski on truth) asserting that the concept of truth is not definable within **PA**: if **T** = **PA** or **T** is extension of **PA**, the set of truth of **T** is not definable in **T**. Denote by $\mathbf{Prov}^{\mathbf{T}}$ the set of provable numbers related to **T** (**T** is an extension of **PA**) and by $\mathbf{Tr}^{\mathbf{T}}$ – the set of true numbers (another label suggested by Tarski; Gödel used a similar terminology). We have:

(**) $\mathbf{Prov}^{\mathbf{T}} \subset \mathbf{Tr}^{\mathbf{T}}$

(**) provided a comparison of the concepts of proof and truth (Gödel was aware of (**), but Tarski exactly proved this fact). Due to incompleteness, the set $\mathbf{Prov}^{\mathbf{T}}$ is not recursive. However, this set is recursively enumerable. Hence, we can (in principle) enumerate the theorems of **PA**, although the set of its non-theorems lacks this property.

But what about the set $\mathbf{Tr}^{\mathbf{T}}$ as far the issue concerns recursiveness, etc.? This collection does not belong to any level of the arithmetical hierarchy (see Murawski 1999, p. 284). Thus, it is not recursively enumerable. And this observation exactly shows how enormous and deep is the real gap between truth and proof. This fact has an additional evidence in the following circumstances. Firstly, **PA** is semantically complete as any first-order theory, because taking all its truths as axioms secures that every arithmetical truth belongs to the set of provable formulas. Of course, this strategy trivializes the problem and, what is more important, introduces an explicit infinistic parameter into metamathematics. Secondly, if we add the rule of ω-induction to **PA**, this theory becomes syntactically complete, but it is so at the expense of non-constructive means (we cannot effectively deal with infinite collections of premises). On the other hand, the theoretical realization of syntactic completeness actually overcomes the first incompleteness theorem, but the class of $\mathbf{Tr}^{\mathbf{T}}$ remains not definable. Thirdly, although both Gödel's incompleteness theorems have effective proofs, Tarski's undefinability theorem does not admit such a proof.

We can simply sum up the preceding considerations by saying that the difference between truth and proof consists in the fact that whereas the latter is finitistic, the former appears as essentially infinistic. We can even generalize this diagnosis to saying that syntax is finitary, but semantics is not. This fact cannot be improved by any imitation of semantics in syntax supplemented by infinitary categories. If we take proof as a criterion of mathematical truth, this difference has a very considerable significance, not only mathematical, but also philosophical. We expect that truth-criteria should be effectively performable. Thus, the slogan "from proof to truth" actually means "if you want to demonstrate that *A* is true, you must prove it". There is no other way (method), because we have no effective procedures in order to review or check all models (all structures in which our statements are true) of a given theory. Yet, semantics shows that although we cannot effectively enter into details of models making our judgments true, we are able to abstract descriptions of it. Clearly, assuming that axioms are provable by fiat, does not solve the situation for ever, although various arguments for the truth of axioms have an importance. Philosophically speaking, the statement "truth is infinistic, proof is finistic" gives a precise wording for the view (already mentioned) that truth transcends proof. Perhaps, this could be repeated with respect to a general question concerning the relation between truth and its criteria, but this issue requires a further discussion.

XII
Constructivism and Metamathematics

Constructivism in the foundations of mathematics comprises a fair variety of views about mathematical theories and their properties. Constructivism is not a uniform proposal in the foundations of mathematics (see Beeson 2004, Bridges, Richman 1997, Troelstra, van Dalen 1988). It comprises – even disregarding some older views as, for instance, Kronecker's account of mathematical reasoning or French semi-intuitionists – many special versions, in particular, intuitionism (as represented by Brouwer and his followers), Russian constructivism, Bishop's constructivism, computable analysis, constructive analysis, predicativism, Lorenzen's operationalism, ultrafinitism (ultra-intuitionism), or constructive type-theory. Every brand of constructivism offers a criterion **C** of constructivity. In my further remarks I will concentrate on intuitionism and Russian constructivism.

Assume that **C** refers to a given fixed criterion of constructivity. Such a criterion always generates a kind of logic considered as constructive. For example, **C** can postulate logic in which the law of excluded middle does not hold – that is, it can postulate the standard intuitionistic logic – or recommend the Markov principle as admissible (the symbol $\neg$ expresses negation, but its meaning depends on the accepted system; in the case of the Markov principle, $\neg$ formalizes classical negation):

(MP) $\forall x(A(x) \vee \neg A(x)) \wedge \neg\forall x\neg A(x) \Rightarrow \exists xA(x)$.[18]

Consequently, a theory **T** is constructive if and only if it satisfies **C**. Typically, constructive systems, as formally based on a kind of constructive logic, are contrasted with theories having classical logic as their logical foundation.

How to compare, evaluate or justify foundational approaches to mathematics? One way consists in taking some philosophical-methodological insights as guiding. For instance, although the Platonist says that a mathematical object exists if and only if it is consistent, the intuitionist claims that consistency forms the necessary condition for existence which must be supplemented by the possibility of construction. This justification is both philosophical and methodological because it concerns existence, the main ontological concept, but, at least in the case of constructivism, it also invokes a method of proving existence;

18 Intuitionism rejects (MP) as not sufficiently constructive.

this means that a methodological (or perhaps even epistemological) parameter clearly enters onto the stage. The opposition between Platonism and intuitionism appears to be crucial. In fact, both philosophical (or foundational) positions exclude each other. The difference between intutionism and formalism is not so transparent, because the latter (at least as represented by Hilbert) is, on the one hand, a quite radical constructivism as admitting finitary metamathematical reasoning only, but, on the other hand, the principle of excluded middle holds in logic accepted by the formalists. Otherwise speaking, whereas formalism favours classical logic as generally correct (it shares this attitude with Platonism), but restricts logical tools used in metamathematics to effective (just finitary) ones, intuitionism assumes its version of constructive logic as suitable in all cases of mathematical reasoning.

No exposition of philosophical foundations of mathematics can abstract from some mathematical ideas. This fact finds its illustration, for instance, in the use of the phrase "finitary mathematical reasoning" as employed in the above characterization of formalism and its relation to intuitionism or in the character of logic accepted by this or that particular view in the foundations of mathematics (logic is a mathematical theory). Yet, formal mathematical ingredients in justification of the foundational views can be more or less explicit, although we should always remember that the border between mathematics and philosophy (in the discussed domain) is somehow vague. In what follows, I would like to point out some problems related to the role of metamathematical properties in justifications of constructivism. I restrict my considerations to intuitionism and Russian constructivism. Regardless of the fact that ordinary or hard mathematicians are usually more or less skeptical about metamathematical investigations, all genuine concepts used in metamathematics have an explicit mathematical content and function in theories which satisfy typical mathematical requirements. In particular, they are definable (or not) by devices used in mathematics, occur in theorems stated under fully specified constraints and provable by definite methods. The domain of such studies is frequently called mathematical foundations of mathematics. Thus, facts derived from metamathematics as a legitimate part of mathematics can (an should) be employed in (possible) justifications of constructivism (and any other declared position in the philosophy of mathematics as well).

If we characterize mathematical theories, we say that they are (or not) consistent, complete, decidable, independent, etc. Not all metamathematical (or metalogical) properties are equally important. Without entering into a discussion about the meaning of importance in this context, we can consider consistency as a mandatory requirement for any mathematical theory and completeness as good everywhere where it is available but required for pure logic (I will incidentally mention other properties as well).[19] Let **T** designate a theory, that is, a set of formulas closed by the consequence operation *Cn* (of course, this operation must be relativized to the assumed version of logic, for instance, classical or constructive). Thus, $Cn\mathbf{T} \subseteq \mathbf{T}$ (since the reverse inclusion is trivial, we have $Cn\mathbf{T} = \mathbf{T}$). The first task consists in investigating which metamathematical properties hold (or do not hold) for **T**. Is it consistent, complete, decidable, etc.? However, except possessing or not some properties, we have the second issue, namely the way of proving whether **T** is consistent or complete. More specifically, the issue concerns the possibility of constructive proofs of such sentences as '**T** is consistent' or '**T** is complete'. In particular, if **T** is constructive, we have a more concrete problem, namely:

(*) Does **T** possess a constructive metatheory (let us denote it by **MT**?

For instance, one can ask: "If **T** is intuitionistic, is **MT** intuitionistic as well?"

The problem posed by (*) is interesting. Clearly, every fact provable constructively (on any understanding of constructivism) is also provable classically. A simple illustration is provided by the relation between direct and indirect proofs. The constructivist does not accept all forms of indirect proofs (details are not relevant here), but the classical position considers this kind of reasoning as legitimate (if formally correct). Consequently, every direct proof can be converted into an indirect proof, although such a manoeuvre rather appears as unproductive, for direct proofs have a better reputation among mathematicians.

The converse problem is much more important. Many theorems have indirect proofs and one can ask whether their transformation into direct reasonings is possible at all. The classical logician considers the issue as open, but adds that if the direct proof is not available, we should accept an indirect one. The constructivist rejects *ad absurdum* all indirect proofs and adds that theorems, which are not proved constructively, cannot be accepted. The most radical constructivist regards such results as meaningless, but a more moderate position recommends waiting for constructive proofs.[20] This latter methodological proposal takes constructivism as a program.

19 Since Jean-Yves Béziau is one of the champions of paraconsistency, I apologize for ignoring paraconsistent logic [this paper appeared in the *Festschrift* for him].

20 This is the picture from the side of ideology. In fact, constructivists sometimes offer non-constructive proofs. Brouwer's fixed-point theorem in topology is a very good

I will argue that we have arguments for regarding the programmatic constructivism as limited. As I noticed earlier, intuitionism and Russian constructivism function as the points of reference. Intuitionism has a distinguished position in the entire constructivist camp. It began the story of the mature constructivism and its logical foundations are particularly well-established. Many logicians considers intuitionist logic as the most important rival of classical logic. Russian (Markov's) constructivism represents a moderate position. (MP) exhibits a clear formal difference holding between both mentioned kinds of the constructive philosophy and metatheory of mathematics. Another interesting distinction between intuitionism and Russian constructivism concerns the Church thesis (CT). Whereas the latter school accepts (CT) in its fully general form (roughly speaking, every computable function is recursive), the former restricts (CT) to the principle that only arithmetical functions are computable (see Trioelstra, van Dalen 1988 for further details). Since CT has its main relevance in developing constructive analysis than in metamathematics, I will not appeal to it in further considerations. I repeat that my main focus consists in examining consistency and completeness in the framework of constructivism (more specifically, within intuitionism). I will do that by deriving some consequences from basic metalogical results, in particular by using the completeness theorem and the first and the second incompleteness theorem.

The completeness property can be defined either syntactically or semantically (see Pogorzelski 1994 for a more comprehensive treatment of metamathematical concepts and results used in this paper). Syntactically speaking, the completeness property is defined by the statement:

(1) A theory **T** is syntactically complete if and only if or any formula A of the language of **T**, $A \in \mathbf{T}$ or $\neg A \in \mathbf{T}$.

If **T** has the property defined by (1), we say that the syntactic completeness theorem holds for it. The syntactic character of this definition (and the related theorem) becomes evident if we observe that only syntactic notions occur in its formulation, because only notions of syntax are involved in it. The other concept of completeness equates the set of theorems (formulas provable by some admitted rules of inference) in **T** and the collection of all sentences of **T** true in every of its models. It is a semantic notion, because it involves the concept of truth. More formally,

---

example in this respect. In general, philosophical ideology is not always coherent with needs of mathematics.

(2) A theory **T** is semantically complete if and only if ($A \in$ **T** if and only if A is true in all models of **T**).

In fact, the completeness property is defined by the if-part of the right part of (2); the only-part of the definiendum expresses the property of soundness. If **T** is semantically complete, then it enjoys the completeness theorem, that is, the statement

(3) If A is a true in**T**, then A is provable in **T**; in symbols, $\models A \Rightarrow \vdash A$.

This proposition is equivalent to:

(4) A theory **T** is consistent if and only if it has a model.

In other words, the Gödel-Malcev completeness theorem (1)–(4) can be applied to arbitrary formal system. However, since higher-order logic plays a minimal role in intuitionism, it is reasonable to limit our further analysis to first-order (that is, elementary) systems. Moreover, the situation of pure logic (denoted by **LOG**) understood as $Cn\varnothing$ and consisting of the propositional calculus (**PC**) plus the quantification calculus (**QC**) is peculiar to some extent. Neither **PC** nor **QC** are syntactically complete; this holds for classical logic as well as for intuitionistic logic. Nonetheless, we have still another concept of completeness, namely the Post-completeness, which is more suitable for logic. Let **LOG** designate a logic. We say that **LOG** is Post-complete if and only if for any formula A of the language of **LOG**, A is a theorem of **LOG** or the set **LOG** $\cup \{A\}$ is inconsistent. Now, **CPC** (the classical propositional calculus) and **IPC** (the intuitionistic propositional calculus) are Post-complete, but **CQC** (the classical quantification calculus) and **IQC** (the intuitionistic quantification calculus) are not. (1) – (4) can be trivially generalized to **LOG**, although the model theory for intuitionistic theories, including logic, is not so uniform as in the classical case; yet I take for granted that intuitionistic semantics is satisfactory. Thus, **LOG** is syntactically incomplete, partially Post-complete, semantically complete and the equivalence of (3) and (4) obtains for it; these conclusions concern $\mathbf{LOG}^{C}$ (classical logic) and $\mathbf{LOG}^{I}$ (intutionistic logic). (3) and (4) and the equivalence of both are valid for extralogical theories. There are extralogical first-order theories (for instance, the theory of linear order) which are syntactically complete, but their expressive power is rather small. Arithmetic of natural numbers, perhaps the most important extralogical mathematical theory is syntactically incomplete (the first incompleteness theorem). The above observations are true with respect to intuitionistic as well as classical systems.[21]

---

21 A difference concerns decidability. **CPC** and **IPC** are decidable. Classical monadic quantification (predicate) logic is decidable, but intuitionistic logic is not. The full **CQC** and **IQC** are undecidable.

Where is the problem? – one may ask. The metamathematical parity of intutionistic and classical systems essentially breaks when the way of proving some results is examined. Clearly, all results concerning classical systems have classical proofs, that is, executed by deductive devices available in **LOG**^{C}. Some of them have also intutionistic proofs, but this fact is not relevant at the present moment.[22] The real problem occurs, when we ask whether all metamathematical theorems valid for the systems based on intuitionistic logic have proofs acceptable from the intuitionistic point of view. No problem is provided by propositional calculus, because we can prove Post-completeness, both the completeness theorem and their equivalence, by means of **LOG**^{I}. The situation becomes radically different in the case of predicate calculus. Since the completeness theorem implies (MP), the intuitionistic proof of the former seems problematic (see Dummett 2000, pp. 154–155), McCarty 1994 for a discussion) as intuitionistically acceptable; recall that the intuitionist does not accept (MP).[23] Of course, the same holds for the completeness property of extralogical theories. Anyway, if a constructive proof of (3) and (4) is to be available even in the case of **LOG**^{I}, constructivism as modelled by intuitionism must be weakened. Yet, the equivalence of (3) and (4) can be intuitionistically established. Since both Gödel's incompleteness theorems have intuitionistic proofs, the completeness theorem in its both versions is actually the critical problem for any constructivism. The role of (MP) shows that Russian constructivism constitutes a very interesting proposal. In particular, the case of the completeness theorem for **IQC** shows that metamathematical arguments are fairly important for understanding what is the content of constructivism as a foundational project. However, the issue of metamathematics in the context of constructivism is not limited to comparing particular versions of constructivism. Incidentally, similar comparisons are also possible between constructivism and other foundational views.

---

22 I will return to the problem of constructive proofs within the classical position at the end of this paper.

23 Some authors (see Swaart 1976, Veldman 1976) claim that they offered purely intuitionistic completeness proofs. However, a deeper discussion (see Troelstra, van Dalen 1988, pp. 729–730, Dragalin 1987, pp. 186–201) demonstrated that the related arguments were based on non-standard semantics and/or concerned so called ⊥–free fragment of intuitionistic predicated calculus. As Michael Dummett says (Dummett 2000, p. 154), the proof via (MP) is "a best possible result" from the constructive point of view. In fact, some proofs of the completeness of **IPC** use classical logic at the level of **MT** (see Dummett 2000, p., 87).

I pass to the problem of consistency by referring to the Peano arithmetic (**PA**; it uses classical logic as its formal skeleton) and the Heyting arithmetic (**HA**; it has the same arithmetical axioms as **PA**, but intuitionsitic logic forms its logical base). Since **PA** is expressible in **HA**, the latter is subjected to incompleteness theorems. In particular, consistency of **HA** can be proved neither in **HA** nor in **PA.** On the other hand, the consistency of **HA** entails the consistency of **PA** (see Dragalin 1987, p, 30, Dummett 2000, p. 24 for the justification of this fact). This fact appears as interesting in itself, because it suggests that the consistency of arithmetic is independent of choosing classical or intutionistic logic as the indented logical base. One could interpret this situation as providing a good point for intutionism, because it suggests that the weakening of classical logic by a rejection of the principle of excluded middle (or other critical logical rule contested by the intuitionist) does not increase the danger of inconsistency of arithmetic.

However, the following reasoning shows that the issue has further aspects, which make the success of intuitionism problematic. The implication (Cons(**PA**) – '**PA** is consistent'; Cons(**HA**) – '**HA** is consistent', $\vdash^{I}$ – intuitionistic provability)

(5) Cons(**HA**) $\Rightarrow$ Cons(**PA**),

can be established intuitionistically (all further steps are intuitionistically sound as well). Thus, (5) can be rewritten as:

(6) $\vdash^{I}$ (Cons(**HA**) $\Rightarrow$ Cons(**PA**)),

Consequently, we a have (via the Bernays provability conditions):

(7) $\vdash^{I}$ Cons(**HA**) $\Rightarrow$ $\vdash^{I}$ Cons(**PA**).

Transposing (7) gives:

(8) $\neg \vdash^{I}$ Cons(**PA**) $\Rightarrow$ $\neg \vdash^{I}$ Cons(**HA**).

Since the proof of consistency of **PA** cannot be carried out in **PA** itself, the consistency of **HA** is unprovable by the devices available in **HA**. The fact (8) poses difficulties for intuitionism, because it shows that the intuitionist is not able to establish intuitionistically the most fundamental property of mathematical theories in which arithmetic is expressible.[24] Thus, there remain faith and empiri-

24 I communicated this conclusion and arguments supporting it to Ann Troelstra. He replied that I am not right because, since the intuitionists do not ascribe a major importance to formalization, they are not particularly sensitive to the argumentative

cal arguments, similarly as in the case of the classical position. Although they might be considered as more or less convincing, it is difficult to consider them as constructively coherent with intuitionism. Incidentally, and pace intuitionism, (5) suggests a relative proof of Cons(**PA**) modulo Cons(**PA**). It can be viewed as an advantage of intuitionism, because if all arguments for the consistency of a weaker theory become arguments for the consistency of a stronger theory, our faith in the latter could be considered as better grounded. Yet, it is still an act of faith based on empirical evidence, but not a rigorous mathematical reasoning.

Three general conclusions can be derived from the above analysis. Firstly, there are various degrees of constructivism in metamathematics. Due to (MP), Russian constructivism is less radical, but more flexible than intuitionism; similar comparisons are possible with respect to the whole constructivist camp. Secondly, given a fixed **C**, a universal constructive **MT** relative to **C** seems improbable, because all known informal treatments, like forwarded in this paper, use classical logic and this situation seems to be fairly stable. Thirdly, constructive arguments are to be preferred even in the domain of classical metamathematics, similarly as in the case of ordinary mathematics, quite independently of a particular view in the foundations of mathematics. It can be illustrated by the fact that the equivalence of (3) and (4) is constructively provable, although both completeness theorems have classical or less constructive proofs, by the availability of the proof of (3) via (MP), and by (5). Some radical construcitivists, for example, Brouwer, argue that non-constructive mathematics is meaningless. Consequently, they cannot say, for instance, that intuitionistic logic is a part of classical logic. Of course, the classicist has no problem with that assertion, because it is known that $\mathbf{LOG}^{\mathrm{C}}$ is the only maximal extension of $\mathbf{LOG}^{\mathrm{I}}$. Consequently, it is possible to develop a detailed account of intuitionism in the classical framework (see Dragalin 1987). The intuitionist can eventually employ the technique of negative translation and replace every classical assertion A by the formula $\neg\neg A$, where the symbol $\neg$ refers to intutionistic negation. However, this move does not help the intuitionist very much, because it leads to negative statements of the type 'it is absurd that A is absurd'. Since, according to intuitionism, the content of $\neg\neg A$ should be generally considered as different from the content of A – provided that the assertion A is established classically, but not constructively – there are no universal tools for comparing the content of arbi-

---

force of metamathematical questions. However, I consider this answer as begging the question. In fact, metamathematics plays a considerable role in Troeltra, van Dalen 1988.

trary statements of ordinary mathematics from the intuitionistic point of view. On the other hand, the classicist can distinguish various degrees of constructivism (this way out was suggested by Andrzej Mostowski in Poland in the 1950s; see Mostowski 1957, Mostowski 1965, p. 106) and look for constructive proofs of various theorems without violating own theoretical position. This means that the program of constructive foundations of mathematics does not require the radical constructivism and its reach should be regarded as an empirical question, which cannot by decided *a priori*.

XIII
Rule-Following and Logic

1. Introduction

If you type the phrase 'rule-following problem' in Google and click, more than 66 millions of results is displayed. Of course, not all are related to the question formulated by Wittgenstein several years ago and widely popularized by Kripke in recent times. I did not calculate the number of items in the Google-list related to the views of Kripkenstein and their variations. Anyway, few first pages are almost entirely related to Wittgenstein and Kripke. What is fairly surprising is that we do not find too many attempts to clarify the concept of rule in the context of rule-following. Most authors take for granted that normativity and correctness are involved in thinking about rules as followed. Although this intuition is certainly right, its insufficiency for a satisfactory account of rule-following immediately turns out to be obvious. The first part of this paper offers an analysis of rule-following based on a more complex foundation. In particular, some devices taken from deontic logic are employed in my proposed solution of the issue. The second part considers the problem of rules of pure logic and their following in making inferences.

2. The concept of rule-following and Wittgenstein's paradox

As it is customary, I will start with Wittgenstein's formulation of the paradox concerning rule-following. Wittgenstein says (see Wittgenstein 1953, sections 201, 202, 203, pp. 81^{e} 82^{e}):

> This was our paradox: no course of action could be determined by a rule, because every course of action can be made out to accord with the rule. The answer was: if everything can be made out to accord with the rule, then it can also be made out to the conflict with it. And so there would be neither accord nor conflict here.
> It can be seen that there is a misunderstanding here from the mere fact that in the course of our argument we give one interpretation after another; as if one is contended us at least for a moment, until we thought of yet another standing behind it. What this shews is that there is a way of grasping a rule which is *not an interpretation*, but which is exhibited in what call "obeying the rule" and "going against it" in actual cases.
> [...]

And hence also 'obeying a rule' is a practice. And to *thing* one is obeying a rule is not obey a rule. Hence it is not possible to obey a rule 'privately': otherwise thinking one was obeying a rule would be the same thing as obeying it.

[...]

Following a rule is analogous to obeying an order.

I do not intend to give a new interpretation of Wittgenstein. In fact, I think that it is impossible to give a coherent interpretation of the quoted passages.

Let us take seriously Wittgenstein's declaration that following a rule is analogous to obeying an order. This immediately suggests its reading as 'following a rule is analogous to obeying an imperative'. Now, imperatives are expressed by sentences of the form OA (reading: it is obligatory that A, A is obligatory; I will use the latter phrase for verbal expressing of deontic sentences, because it is more coherent with ordinary parlance). Furthermore, according to standard deontic logic, we have $FA \Leftrightarrow O\neg A$ (reading: A is prohibited if and only if not-A is obligatory), $PA \Leftrightarrow \neg FA$ (reading: A is permitted if and only if A is not prohibited), and $IA \Leftrightarrow PA \wedge P\neg A$ (reading: A is indifferent if and only if A permitted and $\neg A$ is permitted). Let **X** be a consistent (this assumption excludes trivial cases of normative regulations) set of initial obligations (that is induced by imperatives captured by a suitable set of O-sentences). A normative system **NS** is defined as the set $Cn\mathbf{X}$, relatively to a given set **X** (in fact, we can work with initial obligations, because their consequences do not increase the content of **NS**). Every non-empty set **X** divides the entire universe of all possible actions into three mutually exclusive spheres: obligatory ($\mathbf{OB^X}$), prohibited ($\mathbf{PR^X}$) and indifferent ($\mathbf{IN^X}$; I will omit the upper index in further considerations). Formally speaking (**A** refers to an action described by the sentence A) $\mathbf{A} \in \mathbf{OB}$ if and only if $OA \in \mathbf{NX}$; $\mathbf{A} \in \mathbf{PR}$ if and only if $O\neg A \in \mathbf{NX}$; $\mathbf{A} \in \mathbf{IN}$ if and only if $OA \notin \mathbf{NX} \wedge O\neg A \notin \mathbf{NX}$ (this last condition is justified by definitions taken from deontic logic).Take an obligation expressed by OA. What does it mean to follow this obligation? Clearly, to follow OA consists in undertaking the action described by A or a sentence B which is logically equivalent to A. Let A describe an action and C (also a sentence about an action) designate its logical consequence. Observe that, due to the so-called Ross paradox ($OA \Rightarrow O(A \vee B)$), where B is arbitrary), undertaking the action described by C does not mean the action to which A refers is realized, unless A and C are logically equivalent. And, of course, any action described by $\emptyset A$ (or sentence equivalent with it) does not follow the imperative expressed by OA. Using the well-known construction known as the Lindenbaum algebra, we can form the class $[A]$ containing sentences logically equivalent with A, symbolically $B \in [A]$ if and only if $\vdash A \Leftrightarrow B$. Consequently, a suitable Lindenbaum algebra represents actions that follow a given obligation expressed by OA. At the same time, however, every action belonging to **IN** agrees with *OA*, although we cannot assert that it follows *OA*. Yet, such an action does not disobey the imperative *OA*. Note that although indifference and permission are related, the former is stronger than the latter ($IA \Rightarrow PA$), but the converse entailment does not hold.

Taking into account hints and suggestions stemming from deontic logic directly leads to the conclusion that Wittgenstein's paradox of rule-following disappears, because it is not true that every course of action can be made out to accord with the rule in question. If *OA* is a rule (expresses a rule), then every element of [*A*] follows this order and every element of **IN** (I will omit upper indexes indicating references to **X** or **NS**) is coherent with it. Eventually, one can distinguish between direct following (satisfying) *OA* and indirect following, that is, acting in such way that *OA* is not disobeyed and $O\neg A$ is not disobeyed. Otherwise speaking, this distinction qualifies elements of [*A*] as describing actions directly following *OA* and elements of **IN** as indirectly following this imperative. Direct rule-following and indirect rule-following (not disobeying-rules) are both normative concepts but they essentially differ. The normative character of the former is primitive and directly induced by **NS**, but normativity of **IN** obtains this status derivatively. Now, no interpretation can justify Wittgenstein's paradox of rule-following. Interpretation, at least in its legal understanding, contributes to borderlines between **OB**, **PR** and **IN**. The logical aspect of the discussed issue is entirely independent of accepting, interpreting or grasping rules.

Wittgenstein correctly says that obeying a rule is a practice. Consider the following example. I see a person *x* who does not smoke. This observation neither suggests that *x* obeys the rule 'Do not smoke!' nor supports the conclusion that *x* never used to smoke. Similarly, if I see a person *y* smoking in a place in which is not allowed, it neither suggests that *y* disobeys the order 'Do not smoke!' nor supports the conclusion that *y* overlooked information prohibiting smoking in this place. In order to say (by a person *x'*) that *x* obeys (disobeys) a rule *OA*, at least following ingredients are required: (a) there is a rule *OA*; (b) the rule *OA* is grasped by *x*; (c) *x* intentionally follows *OA*; (d) *x'* grasp *O A*; (e) *x'* observes how *x* behaves and correctly ascribes to *x* intentionality to act in the way *A*.[25] Contrary to Wittgenstein, there is nothing incoherent in a private (in fact, his quoting words by no means are easy to understand them) obeying rules, unless 'privately' is synonymous with 'exclusively by thinking'. However, if we as-

25 The ascribing of intentionality (see (c)) can be more complex in the case of abstaining from an action.

sume this equivalence, the issue becomes trivial: performing any action is not the same as thinking that we perform this action. Another problem is much more interesting, namely the question 'Is it possible to follow a rule *OA* without its internalizing?' The first suggestion is that we should distinguish direct following and indirect following. Whereas one can argue that the latter can obtain (not: obtains) without internalizing rule, the former seems to require internalization.[26] Anyway, the points (a) – (e) and the issue of internalization of rules are strongly associated with the question of the role of social practice in obeying rules.

3. Kripke's mathematical example

Saul Kripke interprets Wittgenstein's paradox of rule-following by employing a mathematical construction (see Kripke 1982, pp. 8–9). If we add normally 68 and 57, the sum of these numbers is equal to 125. However, define a new function *quus* (Gary Ebbs calls it *plush*; see Ebbs 1997, p. 140), denoted by the symbol $\oplus$. The meaning of $\oplus$ is established by the conditions (i) $x \oplus y = x + y$, if $x, y > 57$; (ii) $x \oplus y = 5$, otherwise.[27] Now, the question appears whether $\oplus$ is 'the function [...] previously meant by +' (Kripke 1982, p. 9). Kripke says (Kripke 1982, pp. 10–11):

> Ordinarily, I suppose that, in computing '68 + 57' as I do, I do not simply make an unjustified leap in the dark. I follow directions I previously gave myself that uniquely determine that in this new instance I should say '125'. What are these directions? By hypothesis, I never explicitly told myself that I should say '125' in this very instance. Nor can I say that I should simply 'do the same thing I always did,' if this means 'compute according to the rule exhibited by my previous examples.' That rule could just as well have been the rule for quaddition (the quus function) as for addition. The idea that in fact quaddition *is* what I meant, that a sudden frenzy I have changed my previous usage, dramatizes the problem.

Appealing to being frenzy is naturally a rhetorical figure. According to Kripke, the problem is that we can invent several definitions of functions pretending to being interpreted as denoted by the symbol x and still to argue that we obey rules of addition, relatively to proposed conventions governing various usages of the phrase 'following rules of addition'. He says that his mathematical example sharpens the problem of various possible interpretations of rules.

---

26 This statement is simplified due to possible exceptions.

27 The point (i) is formulated by the formula $x \oplus y = x + y$, if x, $y > 57$ in the first edition of Kripke 1982. It is an obvious error, because 57 $\oplus$ 68 should give 125 as the result. Further editions corrected this mistake.

Kripke locates his version of Wittgenstein's paradox as pertaining to skepticism about meaning. I will not enter into this issue, except making one general remark below. I will also ignore externalism, internalism, behaviourism, truth, warranted assertibility or realism (see Ebbs 1997 for a comprehensive picture of various links of Wittgenstein's paradox and its Kripkensteinian treatment with various philosophical problems; see also Brożek 2012), because these connections make the issue darker. Like in the case of rule-following in a general logical setting, the problem seems to be much simpler than Kripke suggests. It is true that a person x who performs the old addition operation usually (perhaps not never) asks him- or herself that he or she should answer 125 as the result of adding 68 and 57. The crucial point is that x can always ask himself (herself) or someone else, why $68 + 57 = 125$. The final answer belongs to arithmetic and its standard interpretation. Formally speaking, we have the theory **PA** (Peano arithmetic) and its customary interpretation generating the normal model of **PA**. This codification (or formalization if someone prefers) induces definite rules of calculation for + and other arithmetical operations.

We can change this interpretation, wisely or not, and propose new rules of calculation, for instance, involving non-traditional models.[28] Consider $\oplus$ once again. Truly, it gives 125 as the output for 68 and 57 taken as the input values. However, the meaning of $\oplus$ differs from the sense of + . Although both operations are commutative (in particular, $8 \oplus 7 = 7 \oplus 8 = 5$), but their other properties appear as unclear without further ado. To see it, consider the formula $(8 \times 5) \oplus (7 \times 5)$, where the symbol $\times$ refers to the ordinary multiplication. According to (ii), 5 is the result for these values. Since $(8 \oplus 7) \times 5 = 25$, the *quus* operation is not distributive over multiplication, unless we introduce new constraints on $\oplus$. In general, we have no straightforward answer to the question whether $\oplus$ is distributive over $\times$ or not. Another, and perhaps a more important question, concerns the meaning of + in (i). It seems that Kripke understands this symbol as referring to addition without further qualifications. However, it is problematic. The operation + is defined in **PA** by conditions (a) $0 + n = n$; (b) $m^* + n = (m + n)^*$, where the star operator refers to the successor function. Now, we have $0^* + 56 + 68 = (0 + 56 + 68)^* = 125$ as required, but on the other hand and according to (i) $0^* \oplus 56 \oplus 68 = 5$. Thus, either Kripke should change (a) and (b), which is equivalent to changing the meaning of + as defined in the official arithmetic.

28 I do not speak about standard and non-standard models, because principles governing arithmetic with *quus* are unclear (see above).

Or his definition of *quus* cannot be satisfactory from the mathematical point of view for its vague properties.

Clearly, both (i) and (ii) change the ordinary meaning of addition, regardless of whether + and ⊕ give the same outputs for some selected input values. Kripke asks (see Kripke 1982, p. 9): 'Who is to say that this is not the function I previously meant by +?' Although it is not certain whether Kripke forwards this question in his own name or assumes the position of the skeptic, an answer does not present special difficulties and refers to **PA**. Even if we agree that choosing the standard interpretation of **PA** as official and recommending it as deserving rule-following invokes the social practice of mathematicians and does not exclude other alternatives, this assertion does not suffice to support the thesis that we have here various equipollent (that is, having the same calculable force) rules of addition. We have a good arithmetical theory, which exactly defines +, and univocally generates their properties, on the one hand, and *ad hoc* outlined operation ⊕ without determining its features, on the other. Eventually, and it is the already announced remark, one could appeal to typical arguments of skeptics pointing out that if we normalize a theory **T** and its rules in a language **L**, we must refer to the metalanguage **ML** and its meaning-rules. Thus, although rule-following in **L** finds its justification in rules of **ML**, the latter are not unconditionally settled in the metalanguage. Then we pass to **MML**, **MMML** and further *ad infinitum.* In fact, Kripke alludes to this strategy, but this move adds nothing new to the old-fashioned skepticism, which always maintained that every criterion of correctness (validity, truth, justification, etc.) is plagued by the regress fallacy.

How to locate rules for + and various other possible *quuses* by using respective – that is, generated by arithmetic – regions **OB**, **PR** and **IN**? Accepting **PA** as our official arithmetic, we automatically decide that we should follow its axiomatically stated or derived rules as governing correct calculations employing addition (denoted by +) and other arithmetical operations.[29] In particular, we cannot coherently characterize the spheres **OB**, **PR** and **IN** related to rule-following generated by various *quuses*, unless we prove which newly proposed operations of addition constitute admissible (consistent) alternatives to *plus* within an arithmetic. Correspondingly, all computations determined by rules producing different to those required by **PA** belong to **PR**. However, it is still an incomplete

29 I disregard here the problem of non-standard (in the technical sense) arithmetic, for example, arithmetical theories used in non-standard analysis. I only remark that we can either note that these theories belong to the same arithmetical class or to operate by various spheres of arithmetical **OB**.

answer. Assume that **R** is a rule which is consistent with **PA**, for example, the rule for Å, provided that all unclear points are explained (it can be done, for instance, by introducing Å as a partial arithmetical function). Otherwise speaking, we can introduce various *quuses* in order to execute particular cognitive tasks (for instance, calculating in special areas) or for pedagogical aims of teaching children how to compute inside a definite (usually small) subset of natural numbers, for example, how to add numbers smaller than 68. Yet, the analogy between Wittgenstein and Kripkenstein appears as considerably limited. On the one hand, adding such rules is fairly optional (we can, but there is no duty to make such additions), but, on the other hand, if we decide to supplement our arithmetical machinery by new admissible operations, we create new obligatory regions, possibly different than the **PA**-arithmetical **OB**. I have no simple answer to the question whether we should say here about optional spheres or conditionally obligatory subdomains, but I am inclined to the latter qualification.

4. Logical rules and rule-following

Some psychological inquires – for example, the Wason selection test – suggest that inferential logical rules are not followed by agents performing inferences in this sense that even students of logic are committed to elementary errors in making inferences. Other investigations, in turn, suggest something different. For instance, students usually prefer natural deduction over truth-table method if they are asked to check whether a given formula is tautological or not. This seems justify the thesis that our actual cognitive equipment favours genuine inferential strategies over mechanical procedures. Anyway, accessible empirical data hardly shows the extent to which rules of logic determine the logical rule-following. The only lesson to be derived from various data says that inferential agents sometimes correctly follow logical rules and sometimes fail to do that or that sometimes they perform definitely spontaneous inferences. However, as I will argue, logical rules have some peculiar features as far as the issue concerns their following. These properties are generated by metalogic and are entirely independent of how inferential agents behave.

A typical explanation of the main property of logical inferential rules (**LIR**) is that they are sound, that is, cannot lead from truths (true premises) to falsehoods (false conclusions). If $\mathbf{X} \vdash A$ is a rule with **X** as a set of premises and A as a conclusion, we have

(1) If $\mathbf{X} \vdash A \in \mathbf{LOG}$, then $\neg\Diamond(X \in \mathbf{TRUTH} \wedge B \in \mathbf{FALSE})$.

If we remain within pure logic, **LIR** produce theorems from other theorems, ultimately from logical axioms. If A is a logical theorem, we write

(2) $\vdash A$,

where the symbol $\vdash$ refers to the provability operator. Intuitively speaking, (2) says that A is a logical axiom or a theorem deduced from axioms. Consequently, logical theorems (axioms are a kind of theorems) are said to be inferred from the empty set of assumptions. Thus, the operator $\vdash$ magazines, so to speak, logical inferential machinery.

What is the relation between $\vdash$ and $\models$ ($\models A$ means 'A is a logical truth')? If a **LOG** is complete (I consider this property as a substantial feature of logic), we have

(3) $\vdash A \Leftrightarrow \models A$.

Thus, provability from the empty set of premises is equivalent to universal (logical) validity. Interpret logical rule-following as 'obey rules collected by $\vdash$'. Consider the implication

(4) $\models A \Rightarrow \vdash A$.

The last formula says that if A is universally valid, A generates a rule to be logically followed. A special case of (4) (it is justified by the deduction theorem) is captured by

(5) If $\models (A \Rightarrow B)$, then $A \vdash B$..

Thus, if $A \Rightarrow B$ is universally valid, asserting B belongs to rule-following of asserting A. Observe that $\models (A \Rightarrow B)$ is devoid of any explicit normative content. On the other hand, the concept of logical inference rule has obvious normative connotations. If so, (3) and (5) suggest that either the Hume thesis (ought cannot be inferred from is) is broken in the domain of the logical or the concept of universal validity has a normative content.

How to characterize **OB, PR** and **IN** as generated by logic? One could eventually propose an analogy between logic and arithmetic: rules of logic determine **OB**, their negations (in turn, logical inconsistencies) created **PR**, but we can add various extralogical rules in order to enrich the logical inferential machinery. The issue is more complicated. Since propositional calculus is Post-complete (every formula is either its theorem or added to it produce inconsistency), the spheres **OB** and **PR** exhaust the entire domain related to $\vdash$. Yet we can divide all formulas into universal validities, universal inconsistencies and factual truths (or falsehoods). Thus, logical rule-following in the propositional calculus does not preclude dividing our assertions into **OB**, **PR** and **IN**. If we pass to predicate logic, which is complete but not Post-complete, we can form theories consisting of logical theorems and extralogical assertions (for instance, statements of the form 'there are *n* objects'). Logical validity and being provable from the empty set of premises still coincide, the Hume thesis is broken (or the concept of universal validity is broken), we have theories belonging to **OB** and theories belonging to **IN** (first-order logic extended by adding extralogical assertions) and the tripartite division of all formulas into spheres **OB**, **PR** and **IN.** Thus, we have a quite interesting situation. The propositional calculus possesses the strongest force with respect to rule-following, but its expressive power is the smallest one. First-order logic has a greater expressive power, but the force of its rule-following is smaller than the force of the propositional calculus. Generally speaking, because pure logic does not distinguish any extralogical content, the force of their rule-following is strong. On the other hand, extralogical theories are associated with a smaller force of rule-following. If we could measure how strong is rule-following in particular cases, it we would probably have a good device to treat paradoxes offered by Kripkenstein. Unfortunately in general, but fortunately for philosophers, no such method is available.

XIV
Truth-Makers and Convention T

Since Convention T is a very important ingredient of the semantic theory of truth, every comparison of Tarski's construction with other approaches to the concept of truth must, sooner or later, discuss the equivalence:

(1) *S* is true if and only if *A**,

where *A* is a sentence in an interpreted (this qualification is important, because is dispenses us with worries whether propositions or sentences function as bearers of truth) language **L**, *S* is a name of this sentence and the symbol *A** refers to embedding, for example, via translation, of *A* into a metalanguage **ML**. Convention T requires that any materially correct truth-definition **Df** logically entails every instance of (1), that is, the specialization of this scheme for an arbitrary sentence of **L**; such concrete equivalences are called T–sentences, T–biconditionals or T–equivalences. Note that T–sentences something more than usual material equivalences, because we have that **Df** $\vdash$ *B*, for any *B* being an instance of (1) (see Woleński 2008 for a discussion of this problem). According to Tarski, (1) does not constitute a truth-definition, although it can be considered as a partial one. Take the content of the sentence *A* as the set of all its consequences, formally **Cont**(*A*) = *Cn*({*A*}). Clearly, **Cont(Df)** > *B*. In fact, the content of **Df** exceeds the collection (rather rather the content) of all instantiations of (1), because truth-definitions usually contain elements (expressions) which do not occur in T–sentences, for instance, terms borrowed from set theory, like 'the empty set' or 'sequence' as in the case of the semantic definition of truth. Another argument that (1) cannot serve as an adequate truth-definition is that it holds for falsehoods as well.

Kevin Mulligan, Peter Simons and Barry Smith (MSS for brevity) claim in their seminal paper (see Mulligan, Simons, Smith 1984, pp. 10–11) as follows:

> Putnam [...] has argued that Tarski's theory of truth, through its very innocuousness, its eschewal of 'undesirable' notions, fails to determine the concept it was intended to capture, since the formal characterization still fits if we reinterpret 'true' to mean, for instance, 'warrantedly assertable' and adjust our interpretation of the logical constants accordingly. Putnam's conclusion [...] is that if we want to *account* for truth, Tarski's work needs supplementing with a philosophically non-neutral correspondence theory. If we are right that the Tarskian account neglects precisely the atomic sentences, then its indeterminacy is not surprising. [...]. If as we suggest, the nature of truth is underdetermined by theories like that of Tarski, then an adequate account of truth must include

> considerations which are other than purely semantic in the normally accepted sense. Our suggestion here – a suggestion which is formulated in a realist spirit – is that the way to such a theory lies through direct examination of the link between truth-bearers, the material of logic, and truth-makers, that in the world in virtue of which sentences or propositions are true.

Although MSS modestly call their proposal a mere suggestion, the actual task of introducing the concept of truth-makers seems to be much more ambitious in that it offers at least an outline of a full-blooded theory of truth, in particular, metaphysically grounded. In fact, their paper inaugurated a considerable and hot discussion (see Armstrong 2004, and the papers in Beebee and Dodd 2005, Monnoyer 2007, Loewe and Rami 2009).

Disregarding the details of Putnam's criticism of Tarski (however, see Woleński 2001 for a defence of the semantic definition of truth against Putnam's arguments), I will investigate how MSS's account of truth is related to that of Tarski. More precisely, I will discuss an application, if any, of Convention T when truth-makers are used in an explanation of the concept of truth. Two interpretations of the quoted passage are possible. Firstly, the concept of truth-maker supplements the vocabulary of Tarski's theory. Such a reading seems to assume that although Tarski's account correctly captures very general properties of truth, it requires additional conceptual resources in order for the indeterminacy noted by MSS to disappear. This means that Convention T should be fully preserved. Secondly, introducing truth-makers as a notional device leads to a different truth-definition which is partially, or even entirely, at odds with the semantic account. In this case, however, Convention T can be rejected, modified or preserved in its full form. Each of these three possibilities should be excluded in advance; Tarski himself did not claim that his definition of truth satisfies only the requirement of material adequacy established by Convention T. MSS's literal formulations tend to be closer to the first option, because they explicitly postulate supplementing the typical semantic machinery by 'a philosophically non-neutral correspondence theory' in order to achieve a determination of truth-theory. This could suggest that formal semantics combined with metaphysics (realist in the version of MSS) represented by the concept of truth-makers provides a fully determinate, that is, ontologically involved, theory of truth. Yet, and this circumstance suggests the second interpretation as also possible, MSS characterize 'an adequate account of truth' as a philosophical construction that 'must include considerations […] other than semantic in the normally accepted sense.' The meaning of the context 'considerations which are other than purely semantic in the normally accepted sense' can be understood as referring to the theory radically opposed to that offered by Tarski. I will argue that MSS offer a theory which cannot be regarded as the semantic account of truth supplemented by a certain amount of metaphysics.

Alfred Rami (see Rami 2009, p. 3) proposed the following general characterization of truth-makers theories (I deliberately disregard all appeals to the truth-making relation as a necessary connection). All of them assume the so-called truth-maker principle in the following form:

(2) For every *x*, *x* is true if and only if there is a *y* such that *y* is a truth maker for *x*.

This statement implies:

(3) For every *x*, *x* is true, then *x* has a truth-maker;
(4) For every *x*, if *x* has a truth-maker, then *x* is true.

Implication (2) expresses truth-maker maximalism, but (4) is the principle of truth-maker purism. If we combine (3) and (4), we obtain:

(5) For every *x*, *x* is true if and only if *x* has a truth-maker,

which is a more convenient formulation of (2), at least for my considerations in this paper. In order to neutralize semantic antinomies, (5) should be rewritten as the scheme:

(6) *S* is true if and only if a sentence named by *S* has a truth-maker.

This equivalence can be regarded as generating formulas very close to T–sentences. Perhaps, we can introduce the name 'TM–biconditionals' as a label for instances of (6). Consider the sentence (i) 'snow is white'. Assume that English supplemented by a simple mathematical notation serves as a metalanguage. The related T-equivalence for the sentence in question can be written as:

(7) The sentence 'snow is white' is true if and only if **snow** ∈ **WHITE**,

where the word **WHITE** refers to the set of white entities. The right side of (7) translates the sentence (i) into the chosen metalanguage; this is the language of very elementary algebra of classes supplemented by logical constants and syntactic devices allowing us to form names of sentences belonging to the object language). This translation can even be interpreted as pointing out a truth-maker for the sentence in question. Thus, the affinity between T–equivalences and TM–biconditionals is striking (see Smith and Simon 2007, pp. 80–81 for an opposite view).

Rami further observes that a truth-maker theory does not need to accept both statements (3) and (4). Speaking more precisely, he argues that it is fairly possible to accept truth-maker purism without being committed to truth-maker maximalism. On the other hand, Rami qualifies (4) as an analytic truth. His argument makes use of *reductio ad absurdum*. Assume that *A* has a truth-maker, but it is untrue. If *A* has a truth-maker **tm**, it is true in its virtue. However, by assumption *A* is not true. Thus, *A* is true and untrue, which is impossible. In fact, MSS reject (3) in its full generality (see also Mulligan 2007, Smith and Simon 2007) and replace it by a restricted principle:

(8) $A \vdash \exists \mathbf{tm}(\mathbf{tm} \Vdash A)$,

which can be read 'that *A* is true entails that there is a truth-maker **tm** making *A* true'. The principle (4) has the form:

(9) $\mathbf{tm} \Vdash A \vdash A$,

and its meaning is captured by the statement 'what is made true, is true'. Although (8) and (9) implicitly use (1), this fact is not essential, because one can replace *A* by '*A* is true' or 'it is true that *A*' without making any appeal to T–biconditionals.

If someone accepts the maximalist truth-maker theory as David Armstrong does (see Armstrong 2004), that is, with (3), introducing a surrogate of Convention T (note that Armstrong does not make this step), no major problem arises. Let the symbol **TMT** refer to such a truth-maker theory. We can claim that **TMT** is materially adequate if and only if it entails every instance of (2) or (6), that is, a TM–biconditional for any sentence. The issue looks differently in the case of the MSS account. Denote their theory by **TMT'**. First of all, (8) restricts the set of TM–equivalences to atomic sentences and some other cases, for instance, conjunctions of sentences which are simultaneously made true by the same objects as truth-makers. The restricted TM-scheme is expressed by:

(10) A (is true) $\Leftrightarrow \exists \mathbf{tm}(\mathbf{tm} \Vdash A)$.

Unfortunately, we have no simple way to formulate a counterpart of Convention T for **TMT'**. The problem is that the limitations of (1) are extralogical in character and depend on a tension between the truth-functional (extensional) and the mereological character of truth–makers. Hence, the scope of the existential quantifier in (10) is not precisely established in advance. The situation is even worse, because we do not know which elements of **TMT'** imply the correct TM–equivalences. Perhaps, this situation motivated Barry Smith and Jonathan Simon (see Smith and Simon 2007, p. 97) to arrive at the diagnosis according to which we should not define truth via truth-makers, because this task is simply unrealizable. Although **TMT'** justifies some, mostly very simple or elementary, T–biconditionnals, no generally formulated condition of its material adequacy, similar to Convention T, is available. Hence, **TMT'** cannot be regarded as a metaphysically improved semantic theory of truth. It should be considered as an alternative to Tarski's account.

Finally, I would like to make some remarks about the status of (3) and (4). There is a simple argument that the latter is analytic or even a theorem of (meta) logic, but the former is not. In order to make the argument easier, let me rewrite both formulas as:

(11) $\mathrm{T}A \Rightarrow A$;
(12) $A \Rightarrow \mathrm{T}A$.

Formula (11) (see for example Turner 1990) is frequently adopted as one of the axioms of the logic of truth, when truth operates as a modality (see also Woleński 2004c), but (12) is either rejected for its role in generating the Liar paradox or suitably modified as in (1). However, another motivation for rejecting (12) as universally valid can be given. This motivation is completely independent of the problem of semantic antinomies. Suppose that a three-valued logic, for example Łukasiewicz's logic, functions as the basic system. Take a valuation **v** such that $\mathbf{v}(A) = \frac{1}{2}$. Clearly, the metalogical statement '$\mathbf{v}(A) = \frac{1}{2}$' is true, but $\mathrm{T}A$ is false. This observation shows that the implication $A \Rightarrow \mathrm{T}A$ cannot be considered as a theorem of metalogic, although the formula $\mathrm{T}A \Rightarrow A$ still holds in many-valued logic and its metatheory. We have here a simple analogy with alethic modal logic. The operator T behaves quite analogously to the operator $\Box$ expressing the concept of necessity. Any modal logic admits the formula $\Box A \Rightarrow A$ and rejects the formula $A \Rightarrow \Box A$ as a logical truth. If we accept the implication $\mathrm{T}A \Rightarrow A$ as tautological, but reject the reverse conditional $A \Rightarrow \mathrm{T}A$ as logically valid, the formula $\mathrm{T}A \Leftrightarrow A$ shares the fate of the latter and cannot be considered as a logical theorem.

Nevertheless, we have a way to justify the biconditional $\mathrm{T}A \Leftrightarrow A$. Suppose that we accept the following equivalences:

(13) $\mathrm{F}A \Leftrightarrow \neg\mathrm{T}A$, $\mathrm{T}A \Leftrightarrow \mathrm{T}\neg A$,

which postulates that the falsity, non-truth of A and the truth of not-A are equipollent, (12) becomes acceptable. Otherwise speaking, introducing bivalence legitimizes the full T–scheme as a good theorem of metalogic, provided that devices blocking semantic paradoxes are blocked. Thus, the principle of bivalence is a very important ingredient of the semantic theory of truth. It is quite unclear how TM' is related to bivalence and whether if a many-valued logic were analysed by the conceptual machinery of truth-makers, the intermediate logical values would have their own makers or not; the same question concerns falsity-makers (see Armstrong 2004 for a discussion of falsity-makers). Consider the sentence (ii) 'Tomorrow there will be a sea battle'. Certainly, (ii) has no truth-maker at the present moment, but it will or will not have one tomorrow. Some authors (see Nef 2007) propose abstract truth-makers, but this way out seems to be very expensive (too expensive in my opinion) from the metaphysical point of view; the same concerns Josh Parsons' (see Parsons 2005) ideas connecting truth-makers for statements about past and future events with the realism/anti-realism controversy. If we are not radical indeterminists, assertions about the future can have something like possibility-makers before they become realized or not. Even without introducing many-valued logic, the assertion '(ii) has a possibility-maker' is true, but the statement '(ii) has a truth-maker' is false. I guess that **TMT'** or any other non-maximalist truth-maker theory must be supplemented in order to be able to cope with statements about future. No metaphysically grounded theory of truth can ignore this issue, although purely semantic (model-theoretic) constructions do not need to discuss this question. I am inclined to think, unlike MSS and most authors dealing with truth-makers, that semantics should be seen as, in principle, autonomous with respect to ontology or metaphysics. As a corollary we have that the semantic definition of truth as such does not require any metaphysical or ontological enrichment. Thus, Convention T suffices as the condition of material adequacy as far as the issue concerns the very general properties of truth. On the other hand, nothing prevents making syntheses, realist or not, of semantics and ontology. Truth-makers theories go in this direction.

XV
An Analysis 0f Logical Determinism[30]

1. Introduction

The problem addressed in this paper goes back to Aristotle and his considerations about tomorrow's sea battle. In a famous passage in (*De Interpretatione* 19a 25–30; after (*The Works of Aristotle*, vol. 1: *Categoriae and De Intepretatione*, tr. By E. M. Edghill, Oxford University Press, Oxford 1928), the Stagirite says:

> Everything must be either be or not be, whether in the present or in the future, but it is not always possible to distinguish and state determinately which of these alternatives must necessarily come about.
> Let me illustrate. A sea-fight must take place to-morrow or not, but it is not necessary that it either should not take place to-morrow, neither it is necessary that it should not take place, yet it is necessary that it either should or should not take place to-morrow. Since propositions correspond with facts, it is evident that when in future events there is a real alternative, and a potentiality in contrary directions, the corresponding affirmation and denial have the same character.

These words initiated a considerable and long discussion (for historical information and substantial assessments see Bernstein 1992; Cahn 1967; Gaskin 1995; Hintikka 1977; Jordan 1963; Karpenko 1990; Lucas 1989; Vuellimin 1996; Zagzebski 1991, Rice 2013). Basically, two problems were discussed. The first question was historical and concerned how Aristotle should be interpreted. Did he revise logic or not? His text, literally understood, not only defends the principle of excluded middle but also tells us something about the logical value of future contingents, that is, sentences about accidental events which will happen or not. This problem was noted by the Stoics. Their celebrated Master Argument tried to prove that everything that is possible will happen in the closer or further future. The second problem is systematic and concerns the question of how radical determinism or fatalism is related to some fundamental principles of logic, in particular, to the principle of bivalence. Roughly speaking, radical determin-

30 This paper is based on my talk delivered at the conference *Łukasiewicz in Dublin*, held at University College Dublin, July 1996. Peter Simons also participated in this meeting. Some ideas (related to determinism and logic) employed in this paper appeared in Woleński 2006a.

ism (fatalism) is a view that the future is uniquely determined by the past or all future events are necessitated by the past.

The view that classical logic deductively implies radical determinism (**RD**) is called logical determinism. Łukasiewicz himself did not use this label. More precisely, his view was (see Łukasiewicz 1922; Łukasiewicz 1970; see also Simons 1989 for a discussion of Meinong's treatment of the excluded middle) that the law of excluded middle (**EM**) or the principle of bivalence (**PB**), both interpreted metalogically, entail radical determinism, when we add the principle of causality (**PC**).[31] Łukasiewicz constructed three-valued logic as a logical justification of indeterminism and free-will.[32] His reasoning was this. Take **PB** + **PC**. We have (the sign + replaces 'and' for a while):

(1) $\mathbf{PB} + \mathbf{PC} \vdash \mathbf{RD}$

Now, rejecting **PB** or limiting **PC** results in canceling a sufficient condition for **RD**. In fact, Łukasiewicz did more. He attacked both parts of **PB** + **PC** as the sufficient condition because he introduced objective possibilities and thereby limited **PC**. However, it seems that a more satisfactory picture of the situation discussed by Łukasiewicz would be achieved by accepting the principle:

(2) $\mathbf{PB} + \mathbf{PC} \dashv\vdash \mathbf{RD}$,

that is, assuming that bivalence and the principle of causality are logically equivalent to radical determinism.

The term 'logical determinism' (more strictly its German counterpart, that is, *logische Determinismus*) was introduced by Schlick. He spoke about the paradox of logical determinism, consisting in a conflict between logic and determinism. Schlick's formulation is this (Schlick 1931, p. 202):

> […] the principles of contradiction and excluded middle would not rank as statements about future states-of-affairs if determinism did not prevail. In fact, so Aristotle argued, if determinism is true, and if the future, therefore, is not already laid down and *determined* now, that it seems as if proposition "Event *E* will take place tomorrow' could today be be neither true not false.

This problem was also discussed by Waismann who, besides 'logical determinism', also used the expressions 'logical Predestination' or 'logically preordained'

---

31 See Urchs 1992 for an analysis of this question.

32 Some qualifications are here in order (see Jordan 1963). In particular, the early Łukasiewicz maintained that formal logic and ontology are very closely related. However, he abandoned this position later and denied that logic has any metaphysical consequences.

(Waismann 1956; Waismann 1965, pp. 27–34; Waismann 1976, pp. 59–68); the following quotation is taken from (Waismann 1956, pp. 8–9):

> I shall single out for discussion [of the nature of philosophical problems – J. W.] – the question […] whether the law of excluded middle, when it refers to the statements in the future tense forces us into a sort of logical Predestination. A typical argument is this. If it is true now that I shall do a certain thing tomorrow, say jump into the Thames, then no matter how fiercely I resist, strike out with hands and feet like a madman, when the day comes I cannot help jumping into the water; whereas, if this prediction is false now, then whatever efforts I may make, however many times I may nerve and brace myself, look down at the water and say to myself, 'One, two, three' – it is impossible for me to spring. Yet that the prediction is either true or false is itself a necessary truth, asserted by the law of excluded middle. From this the startling consequence seems to follow its is already now decided what I shall to do tomorrow, that indeed the entire future is somehow fixed, logically preordained.

There is a difference between Schlick and Waismann to the effect which logical principles are responsible for logical determinism. Schlick speaks about the laws of contradiction and excluded middle, but Waismann only of the second. Yet, the latter author (Waismann 1965; Waismann 1976) discussed the problem in the sections entitled "The Timeless Nature of Truth" and "Zeitlosgkeit der Wahrheit". This brings us to the relation between the properties of true sentences related to time and the mentioned principles of logic, that is, **PB**, **EM** and **LC**. Leaving the consideration of this issue to the next section lest me add here that neither Schlick nor Waismann accepted logical determinism. Roughly speaking, both rejected the view that logic applies to the empirical world, Schlick because tautologies are devoid of content, Waismann for a distinction between saying true and making true. Otherwise speaking, logic (Schlick, following one of the principal views of the Vienna Circle) or grammar (Waismann being influenced by Wittgenstein's second philosophy) cannot entail ontological consequences.

2. Some basic intuitions

Let me begin with **PB** and related logical principles (see Woleński 2014 for a more comprehensive analysis). First of all, following Łukasiewicz, one must very sharply distinguish the logical and metalogical versions of the laws of excluded middle and (non)-contradiction (**LC**). **PB** says that every sentence is either true or false. Clearly, it is a metalogical principle and, in order to investigate its relation to **EM** and **LC**, they both must be formulated metalogically as well. Thus, we have 'no sentence is jointly true and false' for **LC** and 'every sentence is true or false' for **EM**. Thus, we obtain:

(2) $\mathbf{PB} \Leftrightarrow \mathbf{LC} \wedge \mathbf{EM}$.

However, if we tacitly assume that we are working in a consistent system, **EM** suffices for arguments. It explains why this last principle is used as basic.

The next problem concerns truth and time. Truth is absolute with respect to time if it is independent of time (Twardowski 1902; I neglect other causes of the relativity of truth pointed out by Twardowski). This means (the principle **AT**):

(3) if a sentence is true at a moment t, it is also true at any other moment t'.

This principle can be divided into two parts:

(3a) (eternality) if a sentence is true at t, it is also true at any $t' \geq t$;
(3b) (sempiternality) if a sentence is true at t, it is also true at any $t' \leq t$;

Thus, the time-absoluteness of truth is the product of eternality and sempiternality.[33]

However, the most preliminary important point concerns sentences in the future tense, that is, about future events, intuitively understood as contingent. Consider

(4) it will happen that A (e. g. 'I will be in Warsaw on September 15, 2014').[34]

I assume that (4) (I take it as a concrete example) contains (or implies) as its essential factor

(5) it is possible that I will be in Warsaw on September 15, 2014.

Since we want to have (5) as expressing a contingent sentence, we add:

(6) it is possible that A and it is possible that not A.

Let the symbol $\blacklozenge A$ means that A is contingent (= not necessary and not impossible), but $\Diamond$ and $\Box$ stands for possibility and necessity, respectively. Thus, the contingency of A is defined by:

(7) $\blacklozenge A =^{df} \Diamond A \wedge \Diamond\neg A$ (or $\neg\Box\neg A \wedge \neg\Box A$).

Accordingly, the sentence (5) has the form $\blacklozenge A$ and is read as 'it is contingent that I will be in Warsaw on September 15, 2014'. (**AT**) for temporal future sentences has the form (t is a parameter related to the present time; since $t' \geq t$, the symbol $A(t')$ expresses that the sentence A concerns a future event):

(8) a sentence of the form 'it is possible that $A(t')$' is true at t iff it is true at any $t'' \neq t$.

---

33 The term 'sempiternal' for the property of truth expressed in (3b) was coined by Peter Simons. The problem of the absoluteness of truth in Polish is informally discussed in Woleński, Simons 1989.

34 In fact, I am writing this fragment on January 11, 2014.

Applying (**AT**) to (5) gives (I list all the details):

(9) (a) the sentence 'it is possible that I will be in Warsaw on September 15, 2014' is true today (= Januray 11, 2014) iff it is true at any $t'' \leq t$ (sempiternality);
(b) the sentence 'it is possible that I will be in Warsaw on September 15, 2014' is true today (= January 11, 2014) iff it is true at any $t'' \geq t$ (eternality);
(c) the sentence 'it is possible that I will not be in Warsaw on September 15, 2014' is true today (= January 11, 2014) iff it is true at any $t'' \leq t$ (sempiternality);
(d) the sentence 'it is possible that I will not be in Warsaw on September 15, 2014' is true today (= January 11, 2014) iff it is true at any $t'' \leq t$ (eternality).[35]

Clearly, the conditions listed in (9) are not at odds with the internal consistency of (9) modulo classical logic, although the conjunction 'I will be in Warsaw on September 15, 2014 and I will not be in Warsaw on September 15, 2014' is contradictory.

Having the basic intuitions specified, we can give the following explanations of logical determinism (**LD**):

(10) (a) **LD** is derivable only by rules of logic:
(b) **EM** implies **LD**;
(c) **PB** implies **LD**;
(d) **AT** implies **LD**;
(e) it is logically impossible that there are any worlds with natural laws and the past congruent with the actual world and yet have futures that diverges from the actual world;
(f) it is logically necessary that worlds with the same past and the same natural laws have the same future;
(g) the class of worlds with the same past is unique, that is, there are no different possible worlds;
(h) the class of possible future worlds = {the future of the actual world};
(i) logic excludes that 'it is possible that A and it is possible that not-*A*' is a factor of 'it will be *A*', where *A* is neither tautological nor contradictory.

Without addressing the question whether (10a) – (10i) are exactly equivalent, let me assume that they are closely related.[36] I will argue that all mentioned versions of **LD** are not plausible. In particular, they are not derivable from logic

35 Leśniewski (Leśniewski 1931) recommended to avoid the phrase 'true at *t*'. According to his view, we should include temporal coordinates into contents of sentences. For instance, "*a* is *b* at *t*' is true' is preferable over "*a* is *b*' is true at *t*'.

36 It is easily seen that not all are mutually equivalent. For example, (10b) is stronger than (10c), because **EM** is a part of **PB**. On the other hand, the role of **LC** in **PB** consists in excluding a trivial derivation (inconsistency entails everything). Thus, **EM** is more important for the discussed problem than **LC**.

itself. However, this does not mean that I opt for logical indeterminism, that is, a negation of **LD** in one of its formulations. I will argue that the issue of determinism/indeterminism is independent of pure logic. Thus, the truth of future contingencies as defined in (9) is consistent with determinism as well as indeterminism. Following early Łukasiewicz, I think that logic is ontologically applicable. Otherwise speaking, logic can be understood as applied ontology. On the other hand, I will argue, pace Schlick and Waismann, that pure logic itself do not entail any ontological consequences. However, this view is independent of regarding tautologies as formulas the empty factual content. I rather agree with Łukasiewicz that **RD** = **PB** + **PC**, but this situation does not force us to change logic. As a by-product of my analysis I will offer a simple argument that the sempiternality of truth is equivalent to its eternality.

3. A modal analysis of determinism

Let me start with an octagonal diagram (**Δ**) of oppositions, which generalizes a well-known traditional logical square used for displaying formal relations holding between the categorical sentences ('every *S* is *P*', 'some *S* are *P*', etc.) or modalized sentences 'it is necessary that *A*', 'it is possible that *A*' and their negated forms (see Woleński 1996; Woleński 2008) for the construction of this diagram and its various philosophical applications. For instance, analogical relations hold for deontic modalities:

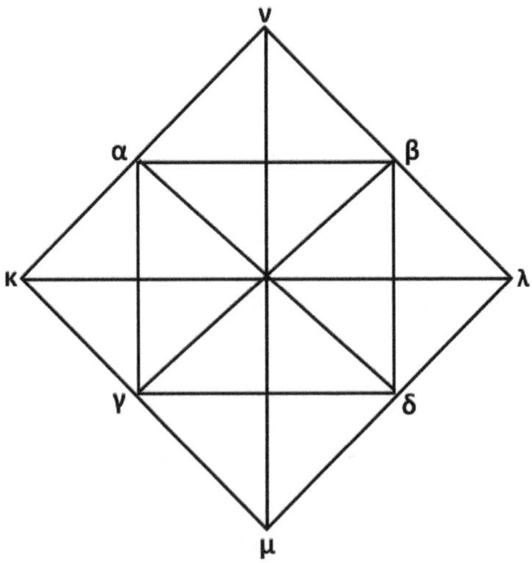

The interpretation of particular points of **Δ**) is as follows: α – *A* is determined (**D***A*); β – ¬*A* is determined (**D**¬*A*); γ – ¬(¬*A* is determined) (¬**D**¬*A*); δ – ¬(*A* is determined) (¬**D***A*); ν – *A* is determined or ¬*A* is determined (α ∨ β; **D***A* ∨ **D**¬*A*); μ – ¬(¬*A* is determined) ∧ ¬(*A* is determined) (γ ∧ δ; ¬**D**¬*A* ∧ ¬**D***A*; *A* is contingent); κ – *A* (it is true that *A*); λ – ¬*A* (it is true that ¬*A*).

We have the following logical dependencies (they constitute the minimal core of modal rules, structurally similar to some theorems of first-order logic):

(11) (a) $\alpha \Rightarrow \varepsilon$;
(b) $\alpha \Rightarrow \gamma$;
(c) $\beta \Rightarrow \varepsilon$;
(d) $\beta \Rightarrow \delta$;
(e) $\neg(\alpha \wedge \beta)$;
(f) $\gamma \vee \delta$;
(g) $\neg(\alpha \Leftrightarrow \delta)$;
(h) $\neg(\beta \Leftrightarrow \gamma)$;
(i) $\mu \Rightarrow \gamma$;
(j) $\mu \Rightarrow \delta$;
(k) $\neg(\nu \Leftrightarrow \mu)$;
(l) $\alpha \vee \beta \vee \mu$;
(m) $\nu \vee \mu$;
(n) $\kappa \Rightarrow \gamma$;
(o) $\lambda \Rightarrow \delta$;
(p) $\kappa \vee \lambda$ ($A \vee \neg A$)
(q) $\alpha \Rightarrow \kappa$ (if *A* is determined, *A* is true)
(r) $\beta \Rightarrow \lambda$ (if ¬*A* is determined, ¬*A* is true).

It is important that the converses of (q) and (r) do not hold. Also the formulas α ∨ β and γ ∧ δ do not express logical principles. We also see that the operator related to γ is not homogenous, because it can be interpreted as α ∨ κ ∨ μ, intuitively '*A* is determined (necessary in a sense) or true or contingent'. From (11) we obtain:

(12) (a) $\mathbf{D}A \Rightarrow \Diamond A$; $\Box A \Rightarrow \Diamond A$;
(b) $\mathbf{D}A \Rightarrow \mathbf{T}A$; $\Box A \Rightarrow \mathbf{T}A$;
(c) $\mathbf{T}A \Rightarrow \Diamond A$;
(d) $\blacklozenge A \Rightarrow \Diamond A$.

The converses of (12a) – (12b) do not hold on purely logical grounds, at least, if **Δ** displays our logic.

Furthermore, we define (recall that *A* is neither tautological nor contradictory):

(13) (a) radical determinism (**RD**): $\forall A(\nu)$, $\forall A(\alpha \vee \beta)$, $\forall A(\mathbf{D}A \vee \mathbf{D}\neg A)$;

(b) radical indeterminism (**RI**): $\forall A(\gamma \wedge \delta)$, $\forall A(\neg\mathbf{D}A \vee \neg\mathbf{D}\neg A)$;

(c) moderate determinism (**MD**): $\exists A(\alpha \vee \beta) \wedge \exists A(\gamma \wedge \delta)$; $\exists A(\mathbf{D}A \vee \mathbf{D}\neg A) \wedge \exists A(\neg\mathbf{D}A \vee \neg\mathbf{D}\neg A)$

(d) moderate indeterminism (**MI**): $\exists((\gamma \wedge \delta) \wedge \exists A(\alpha \vee \beta)$; $\exists A(\neg\mathbf{D}A \wedge \neg\mathbf{D}\neg A) \wedge \exists A(\mathbf{D}A \vee \mathbf{D}\neg A)$;

(e) minimal determinism (**DM**): $\exists A(\alpha)$, $\exists A\, \neg\mathbf{D}A$;

(f) minimal indeterminism (**IM**): $\exists A(\mu)$, $\exists A(\neg\mathbf{D}A \wedge \neg\mathbf{D}\neg A)$;

(g) determinism with respect to a fact f ($\mathbf{D}^{\mathbf{f}}$): $\mathbf{D}f \vee \mathbf{D}\neg f$;

(h) indeterminism with respect to a fact f ($\mathbf{I}^{\mathbf{f}}$): $\neg\mathbf{D}f \wedge \neg\mathbf{D}\neg f$.

Intuitively, **RD** says that everything is determined, **RI** that everything is not determined, **MD** that something is determined and something else is not determined, **MI** that something is not determined and something else is determined, **DM** that at least something is determined, **IM** that at least something is not determined, $\mathbf{D}^{\mathbf{f}}$ that an individually specified fact f is determined, and $\mathbf{I}^{\mathbf{f}}$ that an individually fact f is not determined. **MD** and **MI** are logically indistinguishable, but can differ in the distribution of what is determined and what is not (hence, I changed the succession in the respective formulas, but this has no logical relevance). $\mathbf{D}^{\mathbf{f}}$ entails **DM** and $\mathbf{I}^{\mathbf{f}}$ entails **IM**, **DM** follows from **RD** and **MD**, **IM** follows from **RI** and **MI**.

No statement of (13a) – (13h) expresses a logical truth.[37] This gives a preliminary argument for the thesis that logic and determinism are independent. It can be strengthened by an observation coming from (12). **RD** suggests that $\mathbf{D}A$ and $\Diamond A$ are equivalent. On the other hand, **RI** motivates that $\Diamond A$ and $\blacklozenge A$ are equivalent (modifications for the rest of the possible positions are immediate. Clearly, these equivalences exceed logic. The concept of possibility is in a special position, because $\Diamond A$ is entailed by $\Box A$, $\mathbf{T}A$ and $\blacklozenge A$. Hence, we conclude that the existence of possible events is consistent with **RD**, **RI**, **MD**, **MI**, **DM**, **IM**, $\mathbf{D}^{\mathbf{f}}$ and $\mathbf{I}^{\mathbf{f}}$, that is, with all the possible kinds of determinism and indeterminism listed in (13).

37 In general, I make no specific assumptions about determinism, necessity and related concepts. My arguments work for any mechanism invented for showing that the future is dependent on the past as well as for an absolute or relative understanding of necessity. Additional constraints will be employed on special occasions, for example, when necessity should be interpreted as logical.

4. Logic and determinism

RD is traditionally expressed the sentence 'everything is necessary'. This statement can be formalized as (I will use the box □ instead the letter **D**; I remind that necessity and other modalities can be interpreted ontologically):

(14) $\forall A(\Box A \vee \Box\neg A)$.

In order to prove **LD** by the devices of logic as the only means of inference from a theorem of logic as the only premise, the box should be interpreted as logical necessity. The form of (14) suggests an appeal to **EM** as the starting point of an argument. Thus, we shall adopt this principle, that is, the formula $A \vee \neg A$. Apply necessitation to derive (*) $\Box(A \vee \neg A)$. However, (*) does not imply (14). This fact, very often invoked (for instance, see Wessel 1999, pp. 151–152), does not depend on logical or metalogical understanding of **EM**.[38] Thus, deriving **LD** from (14) is a simple logical mistake, because the latter is not a logical rule. In particular, it does not express **EM**.[39]

Although the above reasoning is simple and nice, it does not solve the question. The argument starts in pure logic and shows that something fails to be derived from it. However, we rather need to show that **LD** does not work in logic or metalogic interpreted ontologically. In order to do that, we should interpret **EM** in terms of **Δ**. Let the symbol ▲ refer to one of □, ◊, **T** or ♦. Schematically speaking, **EM** has the form:

(15) $\forall A(\blacktriangle A \vee \neg\blacktriangle A)$.

with its special instances obtained by replacing the symbol ▲ by one of the members of the set {□, ◊, **T**, ♦}. (15) has a clear ontological import, contrary to the purely logical form, i.e. the formula $A \vee \neg A$. In order to justify **LD**, we should prove that **RD**, that is $\forall A(\Box A \vee \Box\neg A)$, follows from at least one instance of (15). We can skip the formula $\forall A(\Diamond A \vee \neg\Diamond A)$, because it is consistent with all positions and thereby cannot serve as the only basis for **RD**. Also the formula $\forall A(\blacklozenge A \vee \neg\blacklozenge A)$ does not work, because the statement that everything is necessary cannot follow from the sentence that everything is contingent or is not contingent. Eliminating ∀ we obtain $\blacklozenge A \vee \neg\blacklozenge A$. The only route is to apply the proof by cases, but it is blocked by the first disjunct. The same concerns the formula

---

38 The converse entailment holds. The steps from (13) to (*) and from the latter to **EM** are justified by modal logic.

39 This mistake is made in Taylor 1962 (this paper tries to prove fatalism from logic); Rice 2001) uses the label 'logical fatalism'). See Garson 2013, pp. 153–156) for a detailed analysis of Taylor's error.

$\forall A(\Box A \vee \neg\Box A)$, because we need the equivalence $\neg\Box A \Leftrightarrow \Box\neg A$, which begs the question, because it equates non-necessity with impossibility.

To be just, (15) instantiated by $\Box$, $\Diamond$ or $\blacklozenge$ was never used for proving **LD** or questioned, unless **EM** was rejected for other reasons. Thus, there remains:

(16) $\forall A(\mathbf{T}A \vee \neg\mathbf{T}A$,

but we must devote some attention to an interpretation of **T** in the present context before using this form of **EM**. Of course, we cannot understand truth as an epistemological category.

An immediate idea is to read $\mathbf{T}A$ as 'it is real that A'; I will use the symbol $\mathbf{R}A$ for this interpretation and change (16) into:

(17) $\forall A(\mathbf{R}A \vee \mathbf{R}\neg A)$.

Since (12b) and (12c) are fairly plausible for the reading recommended by (17), we can accept that necessity entails reality and that reality entails possibility. The operator **R** behaves differently than $\Box$, $\Diamond$ and $\blacklozenge$, because it admits the equivalence of $\neg\mathbf{R}A$ and $\mathbf{R}\neg A$, which allows us to replace (17) by:

(18) $\forall A(\mathbf{R}A \vee \mathbf{R}\neg A)$

Intuitively, the formula (•) $\neg\mathbf{R}A \Leftrightarrow \mathbf{R}\neg A$ expresses that not being real is equivalent to being unreal jut as 'not true' means 'false'. In fact, (•) can be transformed into the formula (19) $\forall A(\mathbf{R}A \vee \mathbf{R}\neg A) \wedge \neg(\mathbf{R}A \wedge \mathbf{R}\neg A)$. This last formula can be interpreted as the ontological principle of bivalence (**OPB**). It says that everything is real or unreal and nothing is jointly real and unreal. Now we clearly see that neither (18) nor (19) entail **RD** without an additional premise, namely:

(20) $\forall A(\mathbf{R}A \Rightarrow \Box A)$,

that is, without equating reality and necessity. Of course, we can derive various things from $\forall A(\Box A \vee \Box\neg A)$, in particular, $\forall A\Box(A \vee \neg A)$ (see below), $\forall A\Diamond(A \vee \neg A)$ and $\forall A\mathbf{R}(A \vee \neg A)$, but not the formula $\forall A\blacklozenge(A \vee \neg A)$. In fact, if everything is determined, necessities are necessary, possible and real (necessity implies reality and possibility), but not contingent. Yet, these observations give no support for **LD**.

In order to complete the issue of **EM** and its relation to **LD**, we should investigate whether admitting contingencies as occurring ontological items introduces any troubles. Assume that something is contingent, that is:

(21) $\exists A\blacklozenge A$ ($\Leftrightarrow \exists A(\neg\Box\neg A \wedge \neg\Box A)$.

By eliminating $\exists$ from the first formula in brackets, we obtain:

(22) $\neg\Box\neg f \wedge \neg\Box f$.

Define $\neg\Box\neg f$ as $\Diamond f$. This gives:

(23) $\Diamond f \wedge \Diamond\neg f$.

Since $\Diamond A \wedge \Diamond B$ does not entail $\Diamond(A \wedge B)$, we cannot conclude that contradictions are possible, which would threaten **EM** and **PB**. This observation closes the issue of how classical logic is related to determinism. The answer is: in no way. Otherwise speaking, any direct derivation of determinism (the same concerns indeterminism) from pure logic consists in conflating the latter with ontology.

By contrast, let us investigate now what happens when **PB** is limited. According to Łukasiewicz, this step opens the door for indeterminism. If exceptions to **PB** are admitted, for example, some sentences, in particular, future contingents, are valued by the third value **n**, (17) and (18) are not equivalent and become contingent.[40] Let us work with (17). Consequently, its negation (•) $\neg\forall A(\mathbf{R}A \vee \mathbf{R}\neg A)$ is contingent as well. Simple syntactic transformations give (••) $\exists A(\neg\mathbf{R}A \wedge \neg\mathbf{R}\neg A)$ and (•••) $\neg\mathbf{R}f \wedge \neg\mathbf{R}\neg f$, which contradict classical logic, because $\mathbf{R}A$ and $\neg\mathbf{R}\neg f$ are classically equivalent. To save the situation one must change logic. Let f be a future contingency. Thus, both f and $\neg f$ are possible and not necessary as well as unreal. Now, since Łukasiewicz's three-valued logic validates the formula $\Diamond A \wedge \Diamond B \Leftrightarrow \Diamond(A \wedge B)$, we get that $\Diamond(f \wedge \neg f)$. This means that some contradictions are possible. Formally, there is no contradiction, because if f and $\neg f$ have the value **n**, their conjunction $f \wedge \neg f$ has the same status. Hence, the formula $\Diamond(f \wedge \neg f)$ can be true. However, some authors (see Prior 1953) complain about this consequence arguing that contradictions are always false. It seems that Łukasiewicz had a special reason for his solution, namely, that he wanted express contingencies directly in the future tense, without using possibility as an auxiliary device. Nevertheless, there is a special problem with the concept of reality in ontology based on Łukasiewicz' logic. For some reason, he rejected necessary objects, at least in the empirical world, but this was a minor point. Yet, we can ask whether Łukasiewicz's framework admits contingent past events. Assume that $\mathbf{R}f \wedge \Diamond f \wedge \Diamond\neg f$. This entails $\mathbf{R}f \wedge \Diamond(f \wedge \neg f)$. However, this result seems strange when applied to the past, unless we extend the third logical value also to sentences about past contingencies.

40 See Karpenko 1990 for a comprehensive study of the problem of determinism in the context of many-valued logics.

5. Contingencies and classical logic

I begin with the problem discussed at the very end of the last section. Until now, I considered 'being real' as a homogenous predicate. However, something more should be said in order to achieve its full interpretation in the language of classical logic. Due to (12b), the realm of necessity falls under being real, but the former does not exhaust the latter, at least from a purely conceptual point of view. (**Δ**) does not suggest anything about the relation between the real and the contingent. In fact, reality is ambiguous, because it can be determined or contingent. Let us look once more at my visit in Warsaw on September 15, 2014. According to the assumed analysis, it is future and contingent at every time before my arrival to Warsaw. On the other hand, this statement must somehow be supplemented by a qualification of my visit after its realization. Is it real and still contingent? It seems that the following description is adequate (I use the predicative form instead employing operators):

(24) A is real and contingent ($\mathbf{R}^{\mathbf{C}}A$) iff A is real, $\neg A$ is not real, A was possible and $\neg A$ was possible.

Assume that $\mathbf{R}^{\mathbf{C}}A$ holds for some A. Hence, A cannot be necessary and thereby $\Box A$ is false. This suffices for the truth of the implication $\Box A \Rightarrow \mathbf{R}A$ and shows that it is universally valid. The best interpretation of (24) consists in it applying to past contingent events. Thus, the real grows through history.

To accommodate contingencies into classical logic without any problems, I will identify possible worlds with models of maximally consistent sets and use the concept of branchability (Asser 1972, 168–169) as defined by:

(25) (a) a set **X** of formulas branches at a formula A iff the sets $\mathbf{X} \cup \{A\}$ and $\mathbf{X} \cup \{\neg A\}$ are consistent;

(b) a set **X** is branchable iff there is a formula A at which X branches;

(c) a set **X** is branchable iff **X** is a consistent and incomplete set of sentences.

Let **X** be a consistent set of sentences and A be a formula (sentence) independent of **X**. Thus, due to the independence of A with respect to **X,** this set is incomplete and the sets $\mathbf{X'} = \mathbf{X} \cup \{A\}$ and $\mathbf{X''} = \mathbf{X} \cup \{A\}$ are consistent. Consequently, the conditions listed in (25) are fulfilled and **X** branches at A. Note that we do not need to assume that the sets **X'** and **X''** are maximally consistent. We can display this situation by the diagram (**Δ'**)

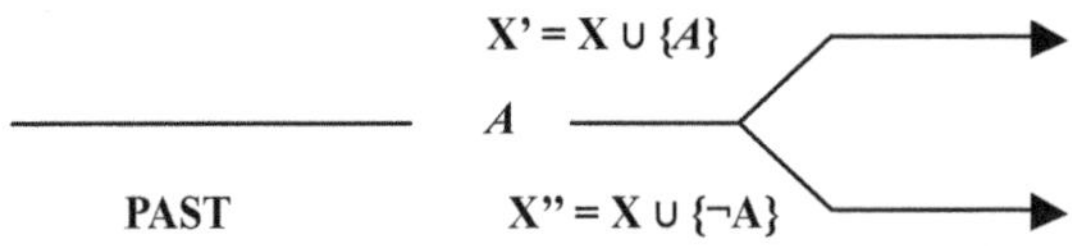

Now we adopt an ontological interpretation.[41] Let A be a sentence uttered at t fixed as present and referring to a future event, and let **X** consist of all true sentences about the past. Since **X** contains only true sentences, it is consistent. It is plausible to assume that A is independent of **X**. If A is independent, **X'** and **X''** are consistent and have models, let us say, **W'** and **W''**, and let **PAST** be the model of **X**. We change the diagram (**Δ'**) into (**Δ''**)

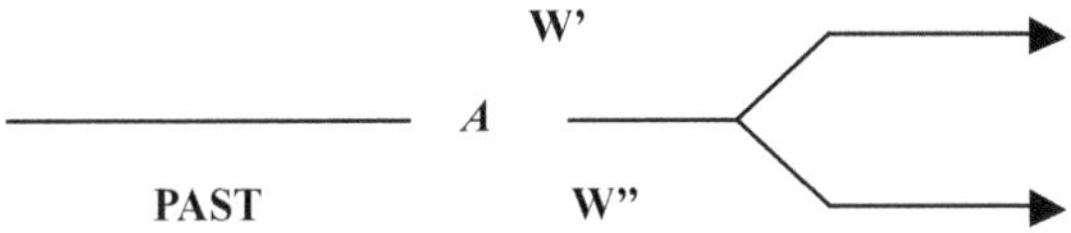

PAST can be considered as the initial segment of **W'** and **W''**. Otherwise speaking, **W'** and **W''** enlarge **PAST**; intuitively **PAST** comprises everything what happened until the moment t (including this moment itself). Moreover, **W'** and **W''** are parts of different possible worlds, that is, models of consistent sets. Let us call them full ontological models. They are different, because the first one, with **W'** as its segment, validates A, but the second, related to **W''** verifies $\neg A$. **PAST** is, of course, the initial segment of the real world $\mathbf{W}^R$, which grows through time. This concurs with an earlier observation (see (24)) that the real contingent is growing. Depending on what will happen, **PAST** will enlarge, at first, to **W'** or **W''**, but then to other worlds, relatively to further branchings.[42]

Now, we are ready to state the following truth-condition for future contingencies. The standard condition for sentences preceded by the operator ◊ says that ◊A is true, A is true at least in one possible world accessible from the world identified as the real one. Consequently, the sentence ♦A is true, if there are two possible worlds **W'** and **W''**, accessible from the real world, such that A is true in one of those worlds, but $\neg A$ in the other. It is enough to fix **PAST** as our real world (or its part).

(26) a sentence 'it is contingent at $t \leq t'$ that $A(t')$' is true in $\mathbf{W}^R$ (= **PAST**) at t iff there are two **W'** and **W''** enlargements of **PAST** such that A is true in **W'** iff $\neg A$ is true in **W''**.

We can also consider possible past scenarios. In particular, (26) can be easily transformed for sentences of the form $\mathbf{R}^C A$ in the following way:

41 I follow Woleński 2012.

42 Needless to say, we consider the relation of **PAST** to a selected future contingent event. This implies nothing about the question of how the future exists: in advance or not, in full detail or not, etc.

(27) $\mathbf{R^C}A(t')$ is true in **PAST** in $t \geq t'$ iff A is true in **PAST**, $\neg A$ is false in **PAST**, but there is a possible world **W'** such that $A(t')$ is true at t'.

Yet, real contingencies can be reduced to future contingencies, because if a sentence is a real contingent, it means that it had to be a future contingency at some moment *t* in the past.

Must we acknowledge that **PAST** is branchable? Of course, we do not need to make this proviso. We can adopt the simplest model, illustrated by the diagram (**Δ'''**)

———————— PAST ———————— *A* ———————— FUTURE ————————▶

This last ontological model displays the main point of **RD**: there is only one course of events. Thus, **RD** is equivalent to the assertion that the **PAST** is not branchable, but branchability automatically introduces a restriction of radical determinism. Finally, radical indeterminism could perhaps be illustrated by a model in which reality consists of separate and unconnected points. Now, **LD** is the view that we can adopt. We can decide on purely logical grounds that (**Δ'''**) offers us a correct description of the world. However, this is fairly impossible, similarly to proving indeterminism (or moderate determinism) by logic. The decision between particular ontological models depends on assumptions concerning the scope and the character of causality and eventually other aspects of regularity, for instance, statistical distributions or fractals. Hence, all models, like (**Δ''**) or (**Δ'''**), are conditional and assert nothing about the actual (meta) physical structure of $\mathbf{W^R}$. Speaking otherwise, logic offers various possible (formal) models for ontology, but it cannot decide which is realized in the world.

Leśniewski (Leśniewski 1912) argued that (3a) and (3b) are logically equivalent, that is, a sentence *A* is eternally true if and only if it is sempiternally true. Leśniewski's original proof is fairly complicated, but it can be very simply reproduced in our framework; I will deal with future contingencies only. Assume that *A* is true at a moment *t.* At first, we will prove that sempiternality entails eternality. If *A* is true sempiternally, it is true at every $t' \leq t$. Since the branching moment is critical, we assume that *t* is just this parameter. Consequently, *A* is true in **W'** or *A* is true in **W''**. Furthermore, *A* is true in **M'** = **PAST** + **W'** or *A* is true in **M''** = **PAST** + **W''**. This implies that $\neg A$ is false in **M'** = **PAST** + **W'** or $\neg A$ is false in **M''** = **PAST** + **W''**. Clearly, *A* cannot change its logical value in a model belonging to {**M'**, **M'**} without producing inconsistency. Thus, if *A* is sempiternally true, it is eternally true as well. To prove the converse implication, suppose that *A* is eternally true, that is, if *A* is true at *t*, it is also true at any moment $t' \geq$ t; of course, *A* is true in **W'** or **W''** and *a fortiori*, in **M'** or **M''**. Consider a moment

$t'' \leq t$. Assume that A is false at t''. This means that A is false in **PAST**. However, this implies that A is false in **M'** or **M''**, contrary to our earlier assumption. Thus, if A is eternally true, it is sempiternally true as well. This result is interesting. Many philosophers, Łukasiewicz is a good example in this respect, accept the eternality of truth as uncontroversial, but complain about sempiternality. The argument for the logical equivalence of (3a) and (3b) shows that something is wrong in this view, because if sempiternality is felt as non-intuitive, the same should concern eternality. It seems that we have here another instance of a conflation of logic and metaphysics. In general, the equivalence of (3a) and (3b) demonstrated that contingent truth is absolute.[43] Since we are working with the concept of truth in a model, we can consider the semantic notion of truth as absolute (Placek 2006). Leibniz proposed the principle *unumquodque quando est, oportet est* (what is, is necessary). However, assuming that the concept of truth satisfies (3), this does not determine that it isnecessary or contingent. Thus, we should rather adopt a weaker principle, namely *unumquodque quando est, reale est* (what is, is real is). Otherwise speaking, what happened could be necessary or contingent. In particular, (6) does not entail that every possibility sooner or later will be realized. This disproves the Master Argument.

43 Note, however, that Łukasiewicz's logic does not force abandoning the that truth is absolute. He could argue that although the third value becomes truth or falsity depending on what will happen, truth (or falsehood) never becomes the third value (or truth). Clearly, the absoluteness of truth (falsehood) is restricted to its eternality. Moreover, this position introduces instability of logical values in general.

XVI
Normativity of Logic

The view that logical rules act as norms or at least performs a normative function was always associated with debates about the nature of logic since antiquity to our times. This fact can be very easily documented by considering various accounts of logic as a science, very sophisticated as well as quite popular. Consider the following list of historical understandings of logic (see Risse 1980); I follow Risse's terminology, but note that the last entry should be phrased by the words *l'art de penser*, if we want to honour the nationality of its proponents of this understanding of logic:

(A) dialectic (analysis and synthesis of concepts; Plato);
(B) analytic (deduction; Aristotle);
(C) *organon* (methods of reasoning; Aristotle);
(D) canonic (norms of knowledge; Epicurus);
(E) *medicina mentis* (descriptive and normative account of mental capacities; Cicero);
(F) *Vernuftslehre* (rules of pure reason; the tradition of *philosophia rationalis*);
(G) *Kunstlehre* (the art of arguing; Husserl);
(H) *Wissenschaftslehre* (the theory of science);
(I) *Denklehre* (the theory of thinking; Arnauld, Nicole).

The labels (C), (D), (E), (G) and (I) have an explicit normative flavour, but (A), (B) and (H) can be easily interpreted as implicitly normative. Take (H), for instance. Petrus Hispanus, a prominent mediaeval representative of the idea of logic as the theory of science nicely expressed this account in the following words: *dialectica* (that is, logic) *est art atrium et scientia scientiarum ad omnium aliarum scientiarum methodorum principia viam habent* (logic is the art of arts and the science of science which establishes methodological principles for all other sciences). Plato offered several recommendations concerning the most reasonable ways of analysis and synthesis of concepts, but Aristotle claimed that any sound scientific method should be based on correct syllogistic moods which constitute the most essential part of the *organon*.

Gottlob Frege had a special account of why logic should be considered as normative. According to him (Frege 1897, p. 128; this paper remained unpublished until 1969):

> Like ethics, logic can also be called a normative science. How must I think in order to reach the goal, truth? We expect logic to give us the answer to this question, but we do not demand that it should go into what is peculiar to each branch of knowledge and

its subject-matter. On the contrary, the task we assign logic is only that of saying what holds with the utmost generality, whatever its subject-matter. We must assume that the rules for our thinking and for our holding something to be true are prescribed by the laws of truth. The former are given along with the latter. Consequently we can also say: logic is the science of the most general laws of truth.

This position has a very strong affinity to Neo-Kantian philosophy. Wilhelm Windelband, the main figure of the Badenian School, characterized formal (*reine*) logic as follows (Windelband 1912, pp. 17–18):

> Die reine Logik oder Logik in engeren Sinne des Worts pflegt wohl als Lehre von der Formen des Denkens definiert zu warden. Doch ist dies zunächst dahin zu spezifizieren, daß es sich in ihr um die *Formen des richtigen Denkens* handelt, die eine durch den Zweck der Wahrheit bestimmte Auswahl aus den psychologisch möglichen Formen der Vorstellungsbewegung darstellen. Es soll in ihr nicht gelehrt warden, wie man wirklich denkt, sondern, wie man *denken soo*, wenn man wirklich denkt will. [...] Alle diese Erwägungern bestimmen endlich den Sinn, in welchem die logische Denkform auch als *Norm* und die formale Logik al seine normative Disziplin zu bezeichnen ist. [...]. [Das] Logische ein "Sollen" [ist]: aber dies Sollen muß seine Grund in etwas haben, dessen Geltung an sich besteht und das erst durch die Beziehung auf ein irresfähiges Bewußtein füdieses zu einer Norm, zu einen Sollen wird.

The similarity of both texts is not surprising. Frege definitely grew in the Neo-Kantian tradition entirely dominant in German academic philosophy in the second half of the 19th century (practically until 1914). Hence, Frege was well acquainted with views similar to that of Windelband. Moreover, Frege attended lectures by Hermann Lotze, a philosopher very close to Neo-Kantianism; in particular, Lotze coined and popularized the term *Geltung* (validity) used by the Neo-Kantians and Frege. The name *die Allgemeingültigkeit* (universal validity), standard in German logical language, comes from *Geltung.*

The Neo-Kantians, following Kant, strongly contrasted two realms, namely that of *Sein* (being) and that of *Sollen* (ougthness). There is no surprise that ethics and aesthetics deal with species of the *Sollen.* Including *das Logische* in the realm of ougthness became a philosophical novelty. Thus, the triad Beauty, Goodness and Truth does not function as in Plato, that is, it does not belong to the world of forms, but constitutes sectors of the *Sollen.* Frege himself (see Frege 1918–1919, p. 351) says that "Just as 'beautiful' points the ways for aesthetics, and 'good' for ethics, so do words like 'true' for logic" (it is likely that Frege was influenced Bruno Bauch, a Neo-Kantian and his colleague from Jena, at this point). This view has various important consequences. In particular, since *Sollen* cannot be derived from *Sein* (it is a counterpart of Hume's thesis about logical separation of is-sentences and ought-sentences), logic is irreducible to psychology. Otherwise speaking, universal validity of logical rules cannot be justified by any psychological data. Anti-psychologism explicitly present in Windelband's treatment of logic, became radicalized in Frege. Another interesting observation points out that the term *Geltung* is (and was) used by many legal philosophers as the main mark of law. The question whether validity of legal norms (orders, prescriptions, etc.) is or not reducible to facts became one of the most basic issues of legal philosophy. Hans Kelsen, the most famous advocate of the separation of *Sein* and *Sollen* in the realm of law, was also a student of the Badenian School.

Still, one additional remark is in order. The normative in Frege's (as well as Neo-Kantian) sense does not refer to practical issues (for instance, Husserl conflated both issues in his idea of *Kunstlehre*). Or, formulating this claim more cautiously, the relation between the normative (as referred to logic and other fields dealing with the *Sollen*) and the practical in logical performances is very far from being clear on Frege's view. On the other hand, such qualifications as *organon*, canonic, *medicina mentis*, *Kunstlehre* or *l'art de penser* immediately suggest that logic has a practical relevance. This point was also strongly indicated by the medieval distinction (see Ebbesen 1980) of *logica docens* (logic as a theory) and *logica utens* (applied logic). One could eventually compare this distinction to the pair consisting of pure mathematics and applied mathematics. Consequently, *logica utens* would employ *logica docens* in actually performed acts of reasoning, for example, similarly to using theoretical principles of arithmetic in practical calculations. Doubtless, there are legitimate cases of applying pure logic in solving some problems, for instance, in constructing programming language. However, one should be very careful in saying that reasoning, including performances intended as instances of deduction, follows principles of *logica docens.* All accessible evidence collected by empirical research (for instance, Wason's selection test) shows that reasoning-performers, even well-versed in *logica docens*, frequently behave at odds with rules of pure logic. In other words, logical 'knowing that' does not force logical 'knowing how'. Moreover, the problem of rule-following (see Kripke 1970) has its modelling also in logic. Although I do not share the radical view that, given a rule **r**, any outcome of behaviour, can be qualified as consistent with **r** (see Woleński 2014a), I consider paradoxes of rule-following as serious and challenging.

Attributing to logic a practical import is tempting. Consider the following declaration of Alfred Tarski (Tarski 1946, p. XV):

> I shall be very happy if this book contributes to the wider diffusion of logical knowledge. The course of historical events has assembled in this country [USA – J. W.] the

most eminent representatives of contemporary logic, and has thus created here especially favourably conditions for development of logical though. These favourable conditions for development of logic can of course, be easily overbalances by other and more powerful factors. It is obvious that the future of logic, as well as of all theoretical science, depends essentially upon normalizing the political and social relations of mankind and thus upon a factor which is beyond of ther control of professional scholars. I have no illusions that the development of logical thought, in particular, will have a very essential effect upon the process of the normalization of human relationships; but I do believe that the wider diffícion of the knowledge of logic may contribute positively to the acceleration of this process. For, on the one hand, by making the meaning of concepts precise and uniform in its own field and by stressing the necessity of such a precision and uniformization in any other domain, logic leads to the possibility of better understanding between those who have the will to do so. And, on the other hand, by perfecting and sharpening the tools of thought, it makes men more critical – and thus makes less likely their being misled by all the pseudo-reasonings to which they are in various parts of the world incessantly exposed today.

Although Tarski was fairly modest in his claims about general profits caused by "the wider diffusion of the knowledge of logic", he believed that the propagation of *logica docens* might strengthen *logica utens* and thereby contribute to improving various political and social relations in the contemporary world.

After digressing about hopes related to social profits caused by logic, let me return to the normativity of logic in a more limited sense. One could expect that this problem, deeply connected epistemology as setting methodological principles for science, disappeared in post-Fregean (that is, mathematical) logic. Although this opinion is true to some extent, normative ingredients are still present in many definitions or characterizations of logic. Here is a sample taken from standard textbooks (subsequently, Russell 1903, p. 10, Church 1956, p. 1, Rosser 1953, p. 5, Kneebone 1963, p. 10, Mendelson 1964, p. 1, Schoenfield 1967, p. 1, Enderton 1972, p. 1, Bonevac 1987, p. 3, Pogorzelski 1994, p. 19, Burris 1998, p. XI, Hinman 2005, p. 1, Nederpelt, Kamarredine 2004, p. VII):

Symbolic or formal logic [...] is the study of the various general types of deduction.
[...] [formal] logic is concerned with the analysis of sentences or of propositions [...] and of proofs [...] with attention to the *form* in abstraction from the *matter*..
The aim of constructing our symbolic logic is that is shall serve as a precise criterion for determining whether or not a given instance of [...] reasoning is correct.
The primary subject-matter of logic is the structure pattern of demonstrative inference.
[...] logic is [...] the analysis of methods of reasoning [...] logic is interested in the form rather than the content of arguments.
Logic is the study of reasoning.
Symbolic logic is a mathematical model of deductive thought.
Logic is the study of correct reasoning.

> Formal logic is [...] a theory of logical consequence.
> Mathematical logic has been, in good part, developed and pursued with the hope of providing practical algorithmic tools for doing reasoning, both in everyday life and mathematics, first by hand or mechanical means, and later by electronic computers.
> Today, modern logic is considered the basis for establishing correctness of theories in mathematics, including mathematical proofs. Logic is also an outstanding scientific tool for the development of correct computer programs.
> One of primary goals of Mathematical Logic is to exploit the advantages bargained by a precise specification of both the language in which mathematics is expressed and the circumstances under which one statements is a consequence of (or follows from) a collection of assumptions.

All quoted explanations (their list could be extended *ad infinitum*) point out, directly or indirectly, that logic studies deductive inferences and some authors supplement their elucidations by adding that logical investigations take into account the form of expressions and abstract from their content.

However, there are differences between the statements quoted above. Some (Russell, Church, Mendelson, Shoenfield, Enderton, Pogorzelski) are purely descriptive and avoid any evaluative or normative qualifications. In turn, Rosser, Kneebone, Bonevac, Nederpelt and Kamarredine, and Hinmann appeal to correctness or preciseness, but Burris practically employs the distinction of *logica docens* and *logica utens* as well as alludes that the former is somehow prior. In fact, we can easily normalize the language by saying that the labels 'deductive reasoning', 'reasoning', 'correct reasoning', 'argument' an 'demonstrative reasoning' are, at least in the domain of mathematics, co-referential or even synonymous, but this move does not liquidate normative ingredients from their content. The issue of preciseness is perhaps different because the expression 'argument' and 'precise argument' are certainly not equivalent in the mathematical parlance. One can, for instance, say that the foundations of calculus were imprecise, but still mathematical (in our sense) in the 17th century and became sufficiently precise in the works of Cauchy and Weierstrass in the 19th century. This, however, means that a test of mathematical preciseness is available and allows us to evaluate what is precise according to some principles and what is not. I will skip the problem of preciseness in my further considerations and concentrate on the normative character of inference rules employed in logic. More specifically, I intend to analyse how the idea of logic as a normative discipline contributes to understanding of the logical (*das Logische* in Frege's sense). To anticipate the conclusion, I will try to show that the normativity of logic is very exceptional as compared to other kinds of being normative and this peculiarity is inherently connected with the universality property of logic (see Woleński

2004a for a more detailed treatment of the universality of logic; I use here some material from this paper). I will assume that classical first-order logic (**FOL** for brevity) is the logic (see Woleński 2004a for a motivation for this choice). I consider **FOL** as first-order logic without identity in order to omit some controversies concerning the status of the identity predicate as a logical constant. Yet, **FO** consists of two segments, namely propositional calculus (**PC**) and predicate logic (**PL**). This fact is important because **PC** and **PL** have significantly different metalogical properties.

We can distinguish four intuitive understanding the universality property attributed to logic, namely

(I) Logic is universal, because it is universally applicable;
(II) Logic is universal, because it is topic-neutral;
(III) Logic is universal, because its principles are universally valid;
(IV) Logic is universal, if it has a rich expressive power.

Since first-order logic possesses a very limited expressive power (not very much extralogical content can be captured by **FOL**), the claim (IV) can be skipped. It does not mean that the question of the expressive power has no general relevance, but my view is that the property to which (IV) refers, belongs to a different level of conceptual analysis than the universality property related to the logical as it is understood in the present essay. The issue becomes fundamentally relevant in the case of the discussion of higher-order logic as the logic or its part. Consequently, one can consider relations between (I) – (III) independently of what is displayed by (IV). A tentative answer is that all items from the set {(I), (II), (III)} are mutually equivalent.

I will justify the suggested answer by those properties of logic which have the exact exposition in metalogic understood as investigations of logical systems via mathematical tools. The first step is to define logic by metalogical devices. Let the definition of logic as a theory of consequence operation serve as the starting point. This account treats logic as a deductive manual. Informally speaking, such manuals instruct how to prove some propositions on the basis of other propositions conceived as premises. It is done with respect to inference rules; for example, modus ponens informs us that it is logically legitimate to infer the conclusion B from A and $A \rightarrow B$ as premises. Assume that **R** is a set of inference rules. The notation $X \vdash^{R} A$ expresses that a formula A is provable (derivable) from the set X of assumptions, relative to the rules of inference from **R** (further I will skip the superscript indexing the provability sign). We define:

(DCn) $A \in CnX \Leftrightarrow X \vdash A$.

Thus, A is a consequence of the set (of propositions) X if and only if it is provable from X.

The consequence operation can be characterized axiomatically. I list below so-called general axioms for Cn plus the deduction theorem and the rule of modus ponens. Adding formulas corresponding to other axioms and/or rules of inference for the propositional calculus and quantifiers of first-order predicate logic completely characterizes FOL, but a general definition of logic does not require this supplement (in fact, some general axioms are also not relevant for defining logic, but they give an idea of what Cn is as a closure operation). The axioms are as follows (card. – the cardinality of the set mentioned in the brackets; L – a language; FIN – finite; $\Rightarrow$ – metalinguistic implication; $\rightarrow$ – implication as an element of L as the object language):

(C1) $\text{card}(\varnothing) \leq \text{card}(\mathbf{L}) \leq \aleph_0$;
(C2) $X \subseteq CnX$;
(C3) $X \subseteq Y \Rightarrow CnX \subseteq CnY$
(C4) $CnCnX = CnX$;
(C5) $A \in CnX \Rightarrow \exists Y \subseteq X \wedge Y \in \mathbf{FIN} \wedge (A \in CnY)$;
(C6) $(A \rightarrow B) \in CnX \Rightarrow B \in Cn(X \cup \{A\})$;
(C7) B Î $Cn(X$ È $\{A\}) \Rightarrow (A$ ® $B)$ Î CnX.

(C1) says that the cardinality of **L** is at most denumerably infinite, (C2) that any set is a subset of the set of its consequences, (C3) established the monotonicity of Cn, (C4) its idempotency, (C5) states the finiteness condition, which means that if something belongs to $Cn(X)$, it belongs to the set of consequences of a finite subset of X (every inference is finitary, that is, performable on the basis of a finite set of premises and, according to the character of rules, finitely long). (C1) – (C5) do not provide any logic in its usual sense, because they do not generate any rules of inference. The fragment of logical machinery is encapsulated by (C6) (the deduction theorem; it has some restrictions in **PL**) and (C7) (modus ponens; it is enough for logic with implication as the sole propositional functor).

How to define logic through Cn? Having the deduction theorem, we say that **LOG** is identified as $Cn\varnothing$. More formally, we have:

(D1) $A \in \mathbf{LOG} \Leftrightarrow A \in Cn\varnothing$, or, equivalently $\mathbf{LOG} = Cn\varnothing$.

(D1) looks artificial or even strange at first sight; clearly, the empty set looks here like a convenient metaphor. In particular – one might argue – we can prove something from the empty set of assumptions just because an amount of logical machinery appears in axioms for Cn. This means that we tacitly assume that (C1) – (C6) display a certain logical content. Hence, the question arises how to justify that stipulations (C1) – (C7) about the consequence operation are proper for logic. As far as the general axioms are concerned, we can, for instance, drop the requirement of monotonicity (it leads to non-monotonic logics used in computer science and belief revision) or finiteness in order to obtain infinitary logics. Hence, any definition of logic through the consequence operation requires an additional justification. (C1) and (C5) are related to human faculties in performing inferences. A possible defence of these axioms consists in pointing out that our inferential performances have the finitary character, because we always employ finite sets of premises of a finite length. This is not at odds with (C1), which admits that the set of sentences can be denumerably infinite, because it means that this set is simply inductively extendible. (C2) is obvious as including axioms and other asserted assumptions among theorems. (C3) says that *Cn*, acting more than once on the given set, produces nothing more. Although the problem of monotonicity is controversial, one can defend this axiom by pointing out that if we can derive something from the empty set, it is also derivable from any other set.

The definition (D1) is equivalent to several other assertions. Among others to ((D2) requires introducing the functor of negation):

(D2) $A \in$ **LOG** if and only if $\neg A$ is inconsistent.

(D3) **LOG** is the only non-empty product of all deductive systems (theories).

(D2) and (D3) define "good" properties of logic. The former qualified negations of logical theorems as contradictions (paraconsistent logicians accept some contradictions as sound, but it is a separate issue), which we expected to be possessed by any reasonable logic (paraconsistency is to be separately discussed at this point). A set X is a theory (deductive system) if and only if $CnX = X$. Thus (D3) states that **LOG** belongs to consequences of every set. Additionally, (D3) (via monotonicity of *Cn*) entails that logical laws are derivable from arbitrary premises. Consequently, if (D2) and (D3) are good, (D1) is too. For our topic, it is important that (D3) somehow generalizes the mentioned statement of Petrus Hispanus that logic provides methodological principles for any scientific field (and for any ordinary argumentation using logical tools). Summing up, (D3) gives a precise account of (I) as an intuition concerning the universal property of logic, that is, its universal applicability. (II) and (III) appeal to semantics, because validity and topicality refer to domains, that is, models. Note that the concept of consequence operation belongs to syntax. Consequently, (D1), (D2) and (D3) define logic syntactically, because they proceed – directly or indirectly – via *Cn*.

The semantic approach to logic suggests:

(D4) $A \in$ **LOG** if and only if for every model **M**, A is true in **M**.

This definition describes logic as consisting of laws (theorems, principles, assertions, etc.), which are true in every model (domain, possible world, interpretation, etc). Now (D4) immediately leads to explicating two other intuitions related to the universality property of logic: it is captured by (II) and (III) and demonstrates that (I), (II) and (III) are equivalent. As I already said, if **LOG** is a part of every theory, it means that it is universally applicable. Exactly the same follows from (D4), because logic as true in every model is applicable in every concrete deductive inference. Further, since **LOG** belongs to every theory **T**, independently of its content, **LOG** is true in every model and **LOG** does not depend on specific assumptions; it is also topic-neutral. Thus, starting from (I) or (II) or (III), we obtain other points, but it is only an informal reasoning. For a formal demonstration. we need to show that (D1) and (D4) are equivalent, that is,

(1) $A \in Cn\varnothing$ if and only if for every model **M**, A is true in **L**.

In fact, (1) is a version of the completeness theorem (CT) (more precisely its weak version), which establishes the parity of syntax and semantics in the area of logic. Another formulation of (CT) is given by ($\models$ – the relation of satisfaction) the statement

(2) for any **M**, **M** $\models A$ if and only if $\varnothing \vdash A$ (or more simply, $\models A$ if and only if $\vdash A$).

To conclude, the universality property of logic as displayed by (III) is formally exposed by (1). This completes the proof that (I), (II) and (III) are mutually equivalent. A nice thing is that the syntactic and semantic approaches to defining logic are equivalent. There is still one interesting result, which contributes to the interpretation of the universality property. **FOL** satisfies the following neutrality theorem (c_i, c_j are individual constants, P_k, P_n are predicate parameters, the notation $A(c)$ and $A(P)$ indicates that a constant c (predicate parameter P) occurs in A):

(N) (a) $A(c_i) \in$ **LOG** $\rightarrow A(c_j/c_i \in$ **LOG**; (b) $A(P_k) \in$ **LOG** $\rightarrow A(P_n/P_k) \in$ **LOG**.

This theorem says that if something is provable in logic about an object or its property (I consider only monadic predicates, that is, referring to properties, but (N) can be easily extended to arbitrary relations), the same can be proved about any other object or property. In other words, logic (more explicitly, **FOL**) does no distinguish any extralogical item, that is, an individual constant or a predicate parameter (individual constants and predicates are not considered as logical objects). Since the (semantic) proof of (N) uses (strong, see below) (CT), this circumstance indicates the link of the former with the universality property and, in particular, topic-neutrality.

The most natural starting point to discuss the normativity of logic consists in taking into account logical rules. The reason is very simple, namely the concept of rule has a normative content by definition. Logical rules are instructions concerning how to derive propositions from other propositions in such a way that inferences are sound (correct). It means that if a rule **R** consists of the premises $A_1, \ldots, A_n$ and the conclusion A (symbolically: $<\{A_1, \ldots, A_n\}, A\}, A>$), it is sound if and only if A is be true provided that the propositions $A_1, \ldots, A_n$ are true. If fact, the soundness property is a part of (CT). More explicitly, it is expressed by the implication asserting that if $A \in Cn\varnothing$, then A is true in all models. Applying logical rules outside pure logic preserves soundness, what is explicitly guaranteed by the strong completeness theorem:

(SCT) for any model $\mathbf{M}$ of the set X, $\mathbf{M} \models A$ if and only if $X \vdash A$.

Soundness can be viewed as a mark of the universal applicability of logic. In fact, it perfectly corresponds with the already quoted words of Petrus Hispanus that logic states methodological principles for all other sciences.

Logic is typically codified by selecting some axioms and rules of inference. A provable logical theorem is an axiom or a proposition inferred from axioms via inference rules. If A is an axiom, it constitutes its own proof. There is also another codification (natural deduction, sequent formalism), which employs inference rules only. Anyway, inferences rules are indispensable in typical codifications. Denote the codification (with axioms or without them) by $\mathbf{LOG}^{\vdash}$. Now (CT) allows to replace provable propositions by universal validities. Denote the set of universally valid propositions by $\mathbf{LOG}^{\models}$. How to compare $\mathbf{LOG}^{\vdash}$ and $\mathbf{LOG}^{\models}$? The issue is simple as far as the matter concerns **PC**. Due to the deduction theorem, we have:

(*) $<\{A_1, \ldots, A_n\}, A> \in \mathbf{LOG}^{\vdash}$ if and only if $(A_1 \Rightarrow (\ldots \Rightarrow (A_n \Rightarrow A)\ldots) \in \mathbf{LOG}^{\vdash}$.

Thus, every logical rule used in **PC** can be reshaped as a logical theorem and vice versa. Moreover, since **PC** is decidable (for instance, by constructing truth-tables), it is not relevant whether we consider theorems (provable formulas) or universal validities (logical tautologies). Yet, $\mathbf{LOG}^{\vdash}$ has a practical advantage, because the collection of universal validities is infinite and cannot be effectively listed. Consequently, $\mathbf{LOG}^{\vdash}$ is the only realistic codification of logical content, not only in **PC**, but also in the entire **FOL**. Although we can, theoretically, skip logical rules and remain with $\mathbf{LOG}^{\models}$, the cost of this way of presenting logic is too high.

The situation radically changes in **PL**. First of all, this theory is not decidable. Hence, although we know (by (CT) that every **PL**-universal validity is provable, we have no algorithm to show whether an arbitrary **PL**-formula is a theorem or not. The second problem concerns the question whether (*) can be extended to **PL** and thereby to the entire **FOL**. The answer is negative because not all **PL**-inference rules can be reformulated as tautologies of **PL** via the deduction theorem. The rule generalization, schematized by the following formula:

(4) $A \vdash \forall x A x$

provides a counterexample, because it cannot be transformed into:

(5) $A \Rightarrow \forall x A x.$

Take '*y* is a natural number' as *A*. It is true, if we substitute 5 for *y*, but the consequent $\forall x A x$ is false. Hence, (5) is not a tautology. This reasoning does not invalidate (4). The expression *Ay* is not a sentence, but an open formula. Consequently, it is neither true nor false. Its intended meaning is captured by the sentence 'any object has the property expressed by *A*'. Intuitively speaking, (4) formalizes the reasoning from any to every, which is sound. Additionally, the discussed example explains why the deduction theorem requires some restrictions in the case of **PL**. More specifically, free variables cannot occur in *A*, if we infer $A \rightarrow B$ from $A \vdash B$. On the other hand, the restriction that *A* must be a sentence, that is, a formula without free variables would decrease the generality of this important theorem.

From a general point of view, inference rules can be divided into derivable and permissible (see Pogorzelski 1994, pp. 111–112, pp. 338–339; see also Appendix at the end of this paper). I will not enter into detailed definitions (quite complicated). Instead, I will limit myself to some general remarks related to **FOL**. If a rule **R** satisfies (*), it is derivable. On the other hand, the rule (4) is permissible (admissible). It cannot be converted into a tautology, but it is sound and does not increase the set of tautologies, although has a relevance for the set of provable formulas. Every derivable rule is admissible (**DER** ⊆ **ADM**), but the reverse statement is not generally true. When the sets **DER** and **ADM** are equal? The answer is that **DER** = **ADM** if and only if a given logical system is Post-complete. Now, a logic **LOG** is Post-complete if and only if adding a non-theorem to **LOG** leads to inconsistency. Since **PC** is Post-complete, the sets $\mathbf{DER}^{\mathbf{PC}}$ and $\mathbf{ADM}^{\mathbf{PC}}$ are equal. On the other hand, **PL** lacks Post-completeness and, thereby, $\mathbf{DER}^{\mathbf{PL}}$ is a proper subset of $\mathbf{ADM}^{\mathbf{PL}}$. Consequently, the entire **FOL** shares the fate of **PL** as far as the issue concerns the relation between $\mathbf{DER}^{\mathbf{FOL}}$ and $\mathbf{ADM}^{\mathbf{FOL}}$.

Although (*) does not hold generally for **PL**, it is not as if it were restricted only to quantifier-translations of **PC**-formulas, for instance,

(6) $\forall xA \wedge \forall xB \Rightarrow \forall xA$ if and only if $\forall xA \wedge \forall xB \vdash \forall xA$

We have, for example, the following equivalences between theorems and rules:

(7) $\forall xA \wedge \forall xB \Leftrightarrow \forall x(A \wedge B)$ if and only if $\forall xA \wedge \forall xB \vdash \forall x(A \wedge B)$, and $\forall x(A \wedge B) \vdash \forall xA \wedge \forall xB$;
(8) $\exists xA \vee \exists xB \Leftrightarrow \exists x(A \vee B)$ if and only if $\exists xA \vee \exists xB \vdash \exists x(A \vee B)$,

which are not reducible do **PC**. Thus, (*) characterizes a large part of the logical, even though it is not strictly universal. Assume now that that a logical rule **R** is an exception with respect to (*). We still have that every formula inferred with the help of this rule is universally valid. This is a trivial consequence of the soundness part of (CT). Consequently, we generalize (*) to the following assertion (the symbol $\vdash^{R}$ expresses provability by **R**):

(**) If $\mathbf{R} \in \mathbf{LOG}^{\vdash}$ it either has a counterpart A such that $A \in \mathbf{LOG}^{\models}$ or for every A, such that $\vdash^{R} A$, $A \in \mathbf{LOG}^{\models}$.

Due to (**), $\mathbf{LOG}^{\vdash}$ and $\mathbf{LOG}^{\models}$ are somehow equivalent. The qualification 'somehow' indicates that this equivalence is not by ($\mathbf{R}^{A}$ – a formula A being a sentential correlate of a rule **R**, like $A \wedge (A \Rightarrow B) \rightarrow B$ is the correlate of the modus ponens):

(9) $\vdash \mathbf{R} \in \mathbf{LOG}^{\vdash}$ if and only if $\mathbf{R}^{A} \in \mathbf{LOG}^{\models}$.

This is so because of the exceptions (see (4) and (5)). Thus, the equivalence in question is rather a consequence of (CT) and the accepted definition of logic. Anyway, (**) establishes a parity of $\mathbf{LOG}^{\vdash}$ and $\mathbf{LOG}^{\models}$.

Soundness is frequently termed as correctness. Both labels are evaluative (in my opinion, 'correct' more than 'sound', but it is perhaps dictated by subjective intuitions). If we say that someone's decision is sound or her behaviour is correct, we presuppose some standards concerning what is sound or correct as contrasted with the unsound or the incorrect. We also say that according to the standards in question, one should or is obliged to undertake sound decisions or to behave correctly and to avoid making unsound decisions or acting in an incorrect manner. In the case of logic, soundness (or correctness) is defined as truth-preservation. Thus, logical rules must preserve truth and qualify inferences in which premises are true and conclusions false (I consider soundness in the case of deductive reasoning; the question of the correctness of inductive or abductive inferences is separate and exceeds the scope of this paper). The use of 'must' in the previous sentence is not accidental. It indicates that the deontic discourse can be applied in the analysis of logical rules. As I have already noticed, they are rules and as such they are normative in their character. Moreover, I suggested that the normativity of logical rules is exceptional as compared to other kinds of normative items and, finally, that this peculiarity is related to the universality property of logic. Now we are sufficiently prepared to a more formal treatment of these informally stated intuitions.

Logical tautologies are true in all models. Consider models as possible worlds. We can ask a question which modal semantics is appropriate for logical truths. Thus, we have the triple (the frame) $<\mathbf{W}, \mathbf{w}^*, \rho>$, where $\mathbf{W}$ is a non-empty set of possible worlds, $\mathbf{w} \in \mathbf{W}$ is the real (actual) world and ρ is a dyadic relation, called the relation of accessibility. In particular, the expression '$\mathbf{w}_1\rho\mathbf{w}_2$' is read 'the world $\mathbf{w}_1$ is accessible from the world $\mathbf{w}_2$' (possibly, $\mathbf{w}_1 = \mathbf{w}^*$ or $\mathbf{w}_2 = \mathbf{w}^*$). The relation ρ can have various properties corresponding to particular modal logic. For example, ρ is not reflexive in deontic logic (I use the term 'logic' in an extended sense in this context). Assume that **o** is an obligation imposed by a legal code (for simplicity, assume that **o** is the only obligation). Deontically speaking, it means that if **o** is valid (in the legal sense) in $\mathbf{w}^*$, every world in which **o** is realized (the sentence describing **o** is true) is a deontic alternative (is deontically accessible from $\mathbf{w}^*$). What about the real world itself? Clearly, it is not a deontic alternative to itself, because it would be unrealistic to expect that people actually realize what should be done. Hence, ρ cannot be reflexive in the case of deontic logic and its semantics. In other words, $\mathbf{w}^*$ is not deontically accessible from itself. Incidentally, this fact justifies Hume's mentioned thesis that is does not imply ought.

Since logic is universally true (true in all possible worlds) including $\mathbf{w}^*$, it does not distinguish any particular world. This means that any possible world is accessible from any other. Thus, the logical ρ has no particular properties. Consequently, the locution 'the world $\mathbf{w}_1$ is accessible from the world $\mathbf{w}_2$' is true in the frame $<\mathbf{W}, \mathbf{w}^*, \rho>$ without any additional constraints. Replace now '*A* universally valid' by '*A* is logically necessary' (symbolically: $\mathbf{N}^L A$). According to the semantics of possible worlds,

(10) '*A* is necessary' is true in a possible world $\mathbf{w}_i$ if and only if *A* is true in every world $\mathbf{w}_k$ such that $\mathbf{w}_k\rho\mathbf{w}_i$.

Informally, a necessity *A* is true in a given world **w** if *A* is true in all worlds which are accessible from **w**. In particular,

(11) $\mathbf{N}^L A$ is true in $\mathbf{w}_i$ if and only if *A* is true in every world $\mathbf{w}_k$ such that $\mathbf{w}_k\rho\mathbf{w}_i$.

Since logical accessibility connects all possible worlds without any additional conditions, $\boldsymbol{\rho}$ is the full relations in the frame <**W**, **w***, ρ>. It means that every pair <$\mathbf{w}_i$, $\mathbf{w}_k$> belongs to $\boldsymbol{\rho}$ (assuming that $\mathbf{w}_i, \mathbf{w}_k \in \mathbf{W}$). (11) validates:

(12) $\mathbf{N}^L A \vdash A$.

This assertion means that logical rules lead from $\mathbf{N}^L$ necessities to truths. However, it is at odds with the deontic semantics, which rejects the formula (**O***A* means '*A* is obligatory')

(13) $\mathbf{O}A \Rightarrow A$ (or $\mathbf{O}A \vdash A$).

The denial of (13) deserves to be called the converse of the Hume thesis. It asserts that is-sentences are not inferable from ought sentences. A more important question, however, concerns the simple Hume thesis that there is an inevitable logical gap between is and ought, which can be formally rendered as:

(14) A not-$\vdash$ **OA**.

However, we have the following assertion:

(15) $A \vdash \mathbf{N}^L A$, provided that A is logically provable or tautological.

The last assertion can be also rendered as:

(16) $\vdash A \vdash \mathbf{N}^L A$ (due to (CT) we have also, $\models A \vdash \mathbf{N}^L A$).

Thus, (16) shows that logical normativity (necessity; I will return to this question in the final part of this paper) does not obey the simple Hume thesis.

One can observe that the violation of Hume's thesis by logical normativity principles results from definitions of logical validity, logical necessity and (CT). I do not deny this comment. However, the task of my analysis concerns the realm of *das Logisches* as a sector of *Sollen.* I claim that the logical normativity is *sui generis* as compared with other kinds of normative phenomena. In turn, Frege's comparison of Beauty, Goodness and Truth was too simplified, because the relation of logic to truth is more complex than Neo-Kantians thought. Assume, as Frege did, that if *A* is true, we should assert *A* and doing so is forced by logic. Everything is OK with this claim, but only if we restrict our game to deductive inferences and logical truths. Since tautologies are true, we should assert them unconditionally. On the other hand, the $\vdash$-machinery, that is, the stock of logical rules, does not force anybody to assert extralogical truts. It seems that Frege confused asserting exclusively on logical grounds with asserting in a more liberal sense. Even if one claims that truths (independently of whether they are logical or not) should be asserted, it does not exclude that errors in his or her assertion acts occur. Although logic can be helpful in reaching truth, its role in this enterprise is limited. This can be shown by a simple reasoning. Assume that Cn closes an assertion, that is, if A is asserted and $B \in Cn\{A\}$, then B is asserted as well. In order to be more realistic, we can add that it is known to the inferring person that $B \in Cn\{A\}$. Moreover, assume the following facts: A is asserted, it is known that $B \in Cn\{A\}$, but B is not asserted. By the deduction theorem, we obtain that $(A \Rightarrow B) \in Cn\varnothing$. Thus, the formula $(A \Rightarrow B)$ is a tautology. Applying the principle of the assertion of tautologies, we obtain that the formula $(A \Rightarrow B)$ is asserted. Since assertion is distributive over implication, we get $(\mathrm{As}(A) \Rightarrow \mathrm{As}(B))$. However, if B is not asserted, A also is not asserted, contrary to the first assumption. Logically speaking, if A is asserted conditionally as a conclusion of a correct inference with asserted premises, it ought to be asserted as well. There is no paradox here. The problem, if any, is caused by a conflation of logical assertions and, say, psychological assertions.

Yet, there is a question of whether logical rules are permissions or instructions. Sometimes, both qualifications are considered as very close in their meaning. To give an example, the modus can be interpreted either as 'it is permitted to infer B from A and $A \Rightarrow B$' or as 'it instructs how to infer B from A and $A \Rightarrow B$'. Both readings agree with the ordinary parlance. On the other hand, permissions are optional, but instructions consist of imperatives. Moreover, we do not need to use all logical inference rules in our deductions, even relatively to the same formalization of logic, which admits various proofs of the same theorems. The possibility of various axioms systems of logic, supplemented by not identical stocks of inference rules, increases the possibility of different proofs of the same tautologies. For instance, **PC** can be formalized with the rule of substitution for propositional variables or without it; the later is used in the present paper. Consequently, if we have a choice between a rule **R** and a rule $\mathbf{R}_1$, treating of inferential means as permissions or alternative instructions seems quite legitimate. However, it is a half of the story. In particular, one should distinguish a question of how to apply logic (this is a problem of *logica utens*) and what is the status of rules as such (this is a question of *logica docens*). I am interested in the latter.

Since I focus on the normative status of logical rules, an appeal to deontic logic seems legitimate. The standard deontic logic frequently adopts the axiom (t is an arbitrary tautology of **FOL**, usually of **PC**):

(17) $\mathbf{O}t$

Roughly speaking, it says that tautologies (tautological acts) obligatory. On the other hand, the formula:

(18) $\neg\mathbf{O}t$

is not a contradiction. It means $\mathbf{P}\neg\boldsymbol{t}$, where **P** is the (weak) permission operator. Take $p \vee \neg p$ as $\boldsymbol{t}$ (it is a typical example considered in presentations of deontic logic). Using the definition $\mathbf{P}A =^{df} \neg\mathbf{O}\neg A$, (18) becomes:

(19) $\mathbf{P}(p \wedge \neg p)$

under this instantiation of $\boldsymbol{t}$. If the world would be completely anarchistic (without any norm establishing obligations), (18) were true. (19) holds in every world in which there is at least one obligation. Thus, (17) says that there is at least one obligation. A fully general deontic axiom should be rather formulated as:

(20) $\mathbf{P}t$.

This formula says that tautologies are permitted. (17) implies (19), but the converse implication does not hold.

Is the deontic case of (19) and (20) applicable to *das Logische*? The negative answer has its motivation in the universal validity of logic. Hence, the division of deontic worlds into the anarchistic and non-anarchistic depends, not on extralogical circumstances, but on an accidental fact that obligations exist or not. Since there is no room for logically anarchistic worlds (that is, models in which logic fails), tautologies are valid in the anarchistic world or worlds (there can be worlds anarchistic from another point than deontic) so literally as in $\boldsymbol{w}^*$. If so, we have reasons to conceive the inference rules, at least satisfying (*), that is, convertible into tautologies as logical obligations. Thereby, a rule $\mathbf{R} = <\{A_1, \ldots, A_n\}, A\}, A>)$ says 'it is obligatory to infer A from the set $\{A_1, \ldots, A_n\}$' (symbolically $\mathbf{O}(A_1, \ldots, A_n, \vdash A)$. A justification for this position can appeal to the fact that since negations of tautologies become contradictions, the same concerns denials of rules. Consider the simplest case, namely:

(21) $\mathbf{O}(A \vdash B)$.

The negation of (21) is:

(22) $\neg\mathbf{O}(A \vdash B)$.

Converting (22) a la (19) gives:

(23) $\mathbf{P}\neg(A \vdash B)$,

and putting the negation sign before B, we obtain:

(24) $\mathbf{P}(A \vdash \neg B)$.

Now we should expect that (24) produces a contradiction. Take an example, namely:

(25) $\mathbf{O}(A \wedge B \vdash A)$.

After transformations, we have:

(26) $\mathbf{P}(A \wedge B \vdash \neg A)$.

Clearly, (26) leads to a contradiction, because it permits to infer $\neg A$ in the situation that the original rule obliged us to infer A. Formally speaking, the sentences $\mathbf{O}A$ and $\mathbf{P}\neg A$ are contradictory. There remain admissible rules, like (4). our reasoning can be repeated also in this case, although it requires a semantic treatment. Anyway, the negation of $\mathbf{O}(A \vdash \forall x A x)$ becomes $\mathbf{P}(A \vdash \exists x \neg A x)$, that is, a contradiction. Note that (22) (and other negated assertions) could not be rendered as $\mathbf{O}(A \neg\vdash B)$, because $A \neg\vdash B$ can refer to derivability of $\neg B$ from A, non-derivability of B from A or independence of B from A (non-derivability of B and non-derivability of $\neg B$ from A).

However, logical rules can be also analysed as permissions. Assume:

(27) $\mathbf{P}(A \wedge B \vdash A)$.

By negating this formula and applying definitions of obligation ($\mathbf{O}A =^{df} \neg\mathbf{P}\neg A$) and prohibition ($\mathbf{F}A =^{df} \neg\mathbf{P}A =^{df} \mathbf{O}\neg A$), we obtain:

(28) $\mathbf{O}\neg(A \wedge B \vdash A)$.

This formula introduces a prohibition of using $\neg(A \wedge B \vdash A$ as a logical rule of inference. A somehow surprising conclusion stemming from the deontic analysis of logical rules of inference is that permission and obligation in the realm of *das Logisches* play the same role. It corresponds with early Wittgenstein's philosophy of logic. According to him, in logic nothing is accidental. If so, it is a small wonder that the assertions 'A is logically necessary', 'A is logically obligatory' and 'A is logically permissible' are equivalent. To strengthen the previous remark, Logic is the only exception to the Hume thesis, because the logical 'is' implies the logical ought and vice versa.

A Discussion

There are several possible objections to my assertions made in this paper. I shall address three points. Firstly, this paper takes **FOL** (with identity) as the logic for granted. I agree that adopting logical pluralism would change the perspective. However, I guess that pluralism is essentially (but not entirely) weakened by two observations: (a) classical logic is the only maximal extension of many alternative logics, in particular, of intutionistic logic; (b) classical logic seems indispensable in metalogic and metamathematics, unless we resign from fundamental results, like (CT) for intuitionistic logic or the Tarski undefinability theorem. Secondly, it can be observed that, contrary to my explanations, logic depends on extralogical facts. For instance, we adopt the definite meaning of logical constants or decide that *Cn* is monotonic. As far as the issue concerns *Cn*, we decide that this operation is defined in such a way that rules of inference preserve truth. However, *das Logische* does not force us to the only understanding of logical constants, the view that *Cn* is monotonic or rules of inference preserve truth. In particular, they can preserve falsity In fact, there are systems of dual or rejection logic with rules preserving being false. As far as the issue of pluralism and introducing an amount of extralogical content into logic via metalogical decisions could influence my deontic analysis of inference rules, the way out consists in replacing obligation and permission by conditional obligation and conditional permission. The trouble is, however, that no entirely satisfactory conditional deontic logic is available so far. Thirdly, one can point out that the treatment of admissible rules as obligations is incorrect, because they should be, as their name suggests, considered as permissions or "optionalities". In fact, admissible rules can be replaced by other admissible rules or even eliminated (see Grzegorczyk 1974, pp. 154–170 for a formalization of **PL** without (4), and with axioms and modus pones as the sole rule of inference). My answer is that this or that formalization is a technical issue and does not change the status of logical theorems and rules. Perhaps, we should contend that although all faithful (that is, preserving the theoremhood) formalizations of **FOL** are equal, a rule **R** is a logical obligation relatively to a given formalization.

XVII
Formal and Informal Aspects of the Semantic Theory of Truth[44]

Tarski's semantic theory of truth (**STT**, for brevity) is usually considered as belonging to metalogic or metamathematics understood as including proof-theory (syntax) and model-theory (formal semantics). This qualification justifies a frequent opinion on **STT** as a formal theory or definition. The title of Tarski 1933, namely *The Concept of Truth in Formalized Languages*, additionally supports such a view.[45] However, a problem appears immediately, when we confront 'formal' in '**STT** as a formal theory' with the list of meanings of this adjective and its contrasts (the second member in every following pair):[46] (a) formal as configuration of elements, structure, form vs. matter, shape, etc.; (b) formal vs. substantial (form vs. content); (c) formal vs. material; (d) formal vs. informal; (e) formal as computable vs. non-computable; (f) formal (effective, finitary, etc.) vs. non-finitary; (g) formal as logical vs. extralogical; (h) formal as syntactic vs. semantic; (i) formal as extensional vs intensional; (j) formal as compositional vs. non-compositional; (k) formal as logical plus mathematical vs. empirical; (l) formal as analytic vs. synthetic; (m) formal as in 'formal language' vs ordinary (informal); (n) *formale* vs. *sachlich* (as in phenomenology). Of course, (a) – (n) are not mutually exclusive and overlap at several points.

Clearly, (a) has no relevance to **STT**. Other oppositions can be implemented to this theory. In order to see it, let me extract several ingredients of **STT**. The

---

44 I use in this paper some material from my publications, namely Woleński 1999a, Woleński (in preparation).

45 Note, however, that the Polish original is titled *Pojęcie prawdy w językach nauk dedukcyjnych* (The Concept of Truth in the Languages of Deductive Sciences). This denomination was changed in German translation of Tarski 1933, published in 1936 (proof-pages appeared in 1935) to *Der Wahrheitsbegriff in den formalisierten Sprachen*. We do not know about motifs of the change, but I guess that Tarski followed a suggestion of the translator (Leopold Blaustein) that the new title would be better for stylistic reasons related to German language. Anyway, the English title is the literal translation of the German one.

46 This list is based on various philosophical dictionaries and encyclopaedias. I did not find any systematic and extensive exposition of the concept of being formal and its antonyms. Typical explanations start from specific cases such as form, formal logic, formal language, etc.

following list, without pretending to be complete, covers the most important (at least according to my view) points: (A) a view about a relation between truth and meaning in **STT** (the truth-definition works for interpreted languages); it is an interpreted language (as understood in **STT**), extensional or intensional; (B) a diagnosis of semantic paradoxes; (C) a way out of semantic paradoxes; (D) the T-scheme; (E) material and formal adequacy (Tarski's own terms) of the truth-definition (in particular the convention T); (F) the condition imposed on metalanguage in order to get a proper truth definition (metalanguage must be essentially richer than object language); (G) the truth-definition itself; (H) bivalence as a consequence of the definition; (I) a theorem that the set of truth for a language **L** is a maximal consistent set; (J) the undefinability theorem (the truth-predicate is not definable in languages sufficiently rich for expressing arithmetic of natural numbers). *Prima facie*, (Ad A) – if truth works for interpreted languages, they are informal as contrasted with entirely formal languages; (Ad B) – informally stated paradoxes, like the Liar antinomy, become formalized in **STT**; (Ad C) – the solution of paradoxes in **STT** proposed by **STT** consists in introducing formal distinctions, in particular, between an object language **L** and its metalanguaga **ML**; (Ad D) – does the T- scheme (*A* is true if and only if *A*; the naive version of T-scheme suffices) generate analytic sentences as its instances?; (Ad E) – the problem of circularity of the definition and its consistency; the status of the claim that the definition must entail all instances of the T-scheme; (Ad F) – is it true, as Tarski claimed, that **ML** in **STT** is syntactic contrary to **L** (a language for which truth is defined)?; (Ad G) – what is relation of satisfaction and truth (perhaps, the latter concept is less formal than the former; does recursiveness of the definition imply compositionality)?; (Ad H) – how to derive the principle of bivalence in **STT** (do we need full classical logic and do we have to fall into circularity, because this logic is bivalent?); (Ad I) is maximality a formal feature or perhaps it has some informal aspects?; (Ad J) – the undefinability theorem shows that the concept of truth is infinitary. Moreover, and according to phenomenology (see (n)), we have (K) – philosophy needs a *sachlich* (substantial, material) theory of truth, not *formale* one. Hence, if **STT** is formal, it has no philosophical relevance or use in philosophical debates related to the concept of truth and, according to phenomenology, this situation cannot be improved.

I cannot deal with all the listed problems and questions. I will entirely omit (A), (B), (C), (D), (E), (H) and (I). I will start with the last problem, namely with the question: Is **STT** a philosophical theory? I studied philosophy in the early 1960s with Roman Ingarden, a student of Husserl. Ingarden used to say at his seminars that **STT** as not *sachlich* is entirely devoid a philosophical significance. Max Black (see Black 1948) argued that Tarski's definition via satisfaction can be added to any known truth-definition, for instance, the coherence theory, the utilitarian theory, etc. Some Tarski's own opinions seem to support such a view. He pointed out in Tarski 1944 that **STT** has not very much to do with philosophy. He declared (Tarski 1944, pp. 361, 362):

> In general, I do not believe that there is such a thing as "the philosophical problem of truth". I do believe that there are various intelligible and interesting (but not necessarily philosophical) problems concerning the notion of truth, but I also believe that they can be exactly formulated and possibly solved only on the basis of a precise conception of this notion (Tarski 1944, p. 361).
> [...] we may accept the semantic conception of truth without giving up any epistemological attitude we may have had: we may remain naive realists, critical realists or idealists, empiricists or metaphycicians — whatever we were before. The semantic conception is completely neutral toward all these issues. (Tarski 1944, p. 362).

Nonetheless, these quotations admit various readings. First of all, Tarski shared the idea of science, according to which several old philosophical problems became scientific when they were precisely formulated. Thus, if someone succeeds in making a given philosophical concept precise, the result becomes scientific. Consequently, the phrase 'the philosophical problem of truth' refers to former philosophical attempts as lacking of a sufficient rigour. Independently of this issue, one should be very careful about deriving too much from Tarski's remarks on naive realism, etc. He probably used adjectives 'naive' and 'critical' in a sense in which they occurred in Polish philosophical writings of the 1930s, that is, for some views concerning perception. My guess is that Tarski wanted to say: whatever view you have about perception, the world naive realism, critical realism or idealism, you can still accept **STT**. Something very similar might be said about a connection of Tarski's theory with the empiricist or metaphysical attitude.

And yet we can easily find quite univocal Tarski's words confirming that he regarded **STT** as a philosophical theory and believed in its philosophical success. Consider the following passages (Tarski 1933, pp. 152, 267):

> The present article is almost wholly devoted to a single problem – *the definition of truth.* Its task is to construct – with reference to a given language – *a materially adequate and formally correct definition of the term 'true sentence'*. This problem [...] belongs to the classical questions of philosophy [...].
> I [...] hope that this work will interest the student of the theory of knowledge above all and that he will be able to analyse the results contained in it critically and to judge their value for further researches in this field, without allowing himself to be discouraged by

> the apparatus of concepts and methods used here, which in places have been difficult and have not hitherto been used in the field in which he works.

Tarski always stressed that **STT** follows the Aristotelian truth-definition and located it inside the correspondence tradition. In particular, he considered the T-scheme as reproducing the idea of the agreement with reality. Doubtlessly, his very modestly expressed hopes that students of epistemology would welcome **STT** as an interesting philosophical proposal became realized beyond all possible expectations and was rationally grounded in 1933. **STT** attracted many leading philosophers, contributed to the semantic revolution, gave rise to several modifications and constructions regarded as alternatives to **STT**, strongly stimulated research in a variety of philosophical problems and, last but not least, found its stable place in textbooks. Even severe criticisms of **STT** provide a considerable evidence that this theory belongs to the centre of contemporary analytic philosophy. These facts entitle us to speak about the Tarski paradigm in contemporary considerations on the concept of truth and semantics.

Presumably all points listed as (A) – (J) have an actual philosophical significance, but I cannot justify this claim in the present paper. Instead, I would like to show the link between **STT** and philosophy by showing that this construction responds to the challenges stated in Russell 1912, as far as the issue concerns correct truth-theories. According to him, any theory aiming to be a successful theory of truth must satisfy three conditions: (i) such a theory should also explain the nature of falsehood; (ii) truth must be taken as a property of beliefs; (iii) truth is an external relation of beliefs to something existing outside them. Disregarding the problem whether (i) – (iii) offer the best collection of constraints imposed on the concept of truth (in fact, they exclude non-relational truth-theories in advance), we can easily show that **STT** directly answers the questions provided by the Russell conditions. Firstly, **STT** explains the concept of falsehood; secondly, **STT** attributes truth to sentences (they can be interpreted as expressing beliefs; and thirdly, **STT** defines truth in model (I use the contemporary account of this theory), that is, relates truth to something external to truth-bearers. Consequently, if we interpret the label 'a philosophical concept of truth' by opposing it to 'a formal theory of truth', **STT** demonstratively falls under the former. Accordingly, Ingarden's qualification of Tarski's construction as formal and thereby useless in philosophy should be rejected.

As I already noticed, the title *The Concept of Truth in Formalized Languages* (or rather its German original) could be adopted by Tarski for stylistic reasons. On the other hand, it (particularly, the qualification 'formalized') provokes a discussion about the application scope of **STT**. Since this theory deals with the context 'true in **L**', the question concerns the domain of the variable **L**. The answer supported by the title of Tarski 1933 says that **L** is formalized. Yet, some commentators intend to replace 'formalized' by 'formal'. This operation leads to reading **STT** as a theory applicable to formal languages. Consider the following quotation (Tarski 1933, p. 166/167):

> It remains perhaps to add that we are not interested here in 'formal' languages in sciences in one special sense of the word 'formal', namely sciences to the signs and expressions of which no material sense is attached. For such sciences the problem here discussed [the problem of truth] has no relevance, it is not even meaningful. We shall always ascribe quite concrete and, for us, intelligible meanings to the signs which occur in the language we shall consider. The expressions which we call sentences still remain sentences after the signs which occur in have been translated into colloquial language. The sentences which are distinguished as axioms seem to us materially true, and in choosing rules of inference we are always guided by the principle that when such rules are applied to true sentences the sentences obtained by their use should also be true.

Tarski uses in this passage the contrast between formal and material in such a way that the former qualification applies to languages devoid of meanings (but not meaningless as absurd) but the latter mark characterizes languages consisting of expressions having definite senses.[47] Thus, a language **L** for which **STT** applies is always interpreted. We can say that if **L** is formal, its full description does not exceed syntax, but material interpretation of **L**, generated by its models, appears as indispensable in the case of doing semantics. Yet, nothing prevents to formalize **L**. To be formalized means that we have to do with the process of formalization and its outcomes. Thus, formalized languages can be interpreted, but formal languages are not interpreted by definition. Otherwise speaking, pure syntax is always formal, but syntax cum semantics (understood referentially or intensionally) has informal features.

The situation described in the previous section becomes clear in the light of the present metalogic. Consider a formalized language **L** for which truth (more precisely, a set of true sentences) is defined. Due to arithmetization (or other similar techniques), the syntax of **L** can be represented in **L** itself. Let us assume that this representation proves that **L** is syntactically formal. However, the

47 I do not claim that Tarski's views about language and meaning have no weak points. In particular, he did not define the concept of meaning. In fact, he intentionally avoided this question and deliberately preferred to speak about interpreted languages as semantic items. For Tarski, the concept of language was clearer than the concept of meaning. See Patterson 2012 for an extensive presentation of Tarski's philosophy of language. Yet **STT** does not seem to be especially dependent on a concrete theory of meaning.

Tarski undefinability theorem shows that semantics of **L** is not fully expressible in **L** itself. In order to define semantic relations, we need to use **ML** (a metalanguage with respect to **L**) which has a greater expressive power than **L**. Perhaps, the most important observation is that **ML** remains partly informal. Even if we formalize **ML**, we must use **MML** and the story reappears. Generally, speaking the hierarchy $\mathbf{L}_1$, $\mathbf{L}_2$ (= $\mathbf{ML}_1$), $\mathbf{L}_3$ (=$\mathbf{ML}_2$), ..., $\mathbf{L}_n$ ($\mathbf{ML}_{n-1}$), ... of languages has the following property: if $\mathbf{L}_k$ is formalized, $\mathbf{L}_{k+1}$ ($1 < k < n$) has some informal features with respect to $\mathbf{L}_{k+1}$ (= $\mathbf{ML}_k$). Thus, there occurs a necessary connection between formal and informal aspects of **STT**. Yet, one point requires an explanation. According to Tarski, the truth-definition for **L** generated by **STT** can be given in the syntax of **ML**. Does it mean that semantics is reducible to syntax? Not at all. Tarski worked in the paradigm of logic on which set theory belonged to logic. In fact, the definition of truth via satisfaction (see below) proceeds in the set-theoretical framework. Thus, we can say at most that defining semantic relation for **L** assumes a set theory. Putting this fact into the contemporary fashion, if **T** is a first-order theory (principally every deductive theory can be expressed in the first-order language), its semantics can be constructed in weak second-arithmetic with the arithmetic axiom of comprehension, employed as a metatheory **MT** of **T**. However, this does not mean that we reduce semantics of **T** to **T**-syntax. Otherwise speaking, if **T** is completely formalized, **MT** does not admit such a treatment.

The semantic definition regards truth as a special case of satisfaction (satisfaction by all sequences of objects). Consider two stocks of ideas. For simplicity, I limit informal explanations for monadic formulas, that is, of the type *P*(...), where an individual variable or an individual name occurs in the place denoted by the symbol (...); the letter **X** represents the assumed universe of discourse.

(1) (General case): open formulas, satisfaction by an object from a given set **X**, non-satisfaction by an object from the complement of **X**;
(Special case): closed formulas (sentences), satisfaction by all objects from a given set **X** (what is this?), satisfaction by no object from a given set **X** (what is this?)

Traditionally, a (declarative) sentence is something that is true or false. Then it is quite natural to insert truth and falsity in the place of question marks in (1). We even have the following proportions:

(2) Open formulas: satisfaction by an objects from a given set **X**;
Sentences: satisfaction by all objects from a given set **X** (truth);
Open formulas: non-satisfaction by the objects from the complement of **X**;
Sentences: satisfaction by no object belonging to **X** (falsity).

Technically, since formulas can be of an arbitrary finite length, it is convenient to introduce infinite sequences of objects in order to generalize (1) and (2) for all possible syntactic cases. Thus, truth is defined – it is an outcome of a very elegant analogy displayed by (2) – as satisfaction of a sentence by all infinite sequences of objects (another equivalent possibilities are satisfaction by some infinite sequences or satisfaction by the empty sentence). One can think, following Tarski to some extent, that this definition, although mathematically nice, has no intuitive content and thereby looks as a formal trick. Perhaps, one could be inclined to think about sequences as facts. However, it is a blind path. The version via the empty sequence entirely clarifies the issue (see Woleński 2003a), because this route expresses a simple circumstance that truth of a sentence has nothing to do with valuation of free variables. It is hard to be surprised by that, because free variables do not occur in sentences (closed formulas). How to express this formal aspect in an informal way? The answer is simpler than one could expect, because truth and falsity of a sentence depends on the very structure of the model in question. Similarly, satisfaction or non-satisfaction of an open formula is determined by how things behave in models and how free variables are valued. Once again, truth appears as a special case of satisfaction.

Very important consequences follow from comparing truth and proof. A deductive proof of a formula A from the premises $\{A_1, A_2, ..., A_n\}$ is a finite sequence of formulas $A_1, A_2, ..., A_n, A_{n+1}$ such that $A_{n+1} = A$ and every step from A_i to A_{i+1} (1 £ i £ n) has a justification in admissible rules of inference (I omit further details concerning the rules). We write $A_1, A_2, ..., A_n \vdash A$, just in the case if A is provable from the assumptions $A_1, A_2, ..., A_n$. Another account of the notion of proof employs the notion of the consequence operation, denoted by the symbol Cn, defined by:

(*) $A_1, A_2, ..., A_n \vdash A$ iff $A \in Cn\{A_1, A_2, ..., A_n\}$.

If $A_1, A_2, ..., A_n \in Cn\mathbf{X}$, we write $A \in Cn\mathbf{X}$. The operation Cn satisfies the so-called general axioms: (α) Cn acts on at most infinitely denumerable languages (that is, having infinitely denumerable number of finitely long formulas): (β) for any set $\mathbf{X}$, $\mathbf{X} \subseteq Cn\mathbf{X}$; (γ) if $\mathbf{X} \subseteq \mathbf{Y}$, then $Cn\mathbf{X} \subseteq Cn\mathbf{Y}$ (monotonicity); (δ) $CnCn\mathbf{X} = Cn\mathbf{X}$ (idempotency); (ε) if $A \in\exists\ Cn\mathbf{X}$, then $\exists\mathbf{Y} \subseteq \mathbf{X}$ $(\mathbf{Y} \in \mathbf{FIN} \wedge (A \in Cn\mathbf{Y})$ (finiteness).[48] A theory $\mathbf{T}$ is a set of sentences closed by Cn (symbolically: $\mathbf{T} = Cn\mathbf{T}$).

48 Other accounts of *Cn* are possible. For instance, we can admit non-denumerable languages, infinitely long formulas, cancel monotonicity, etc. Since *Cn* captures the in-

Assume that a given **T** is expressed in a language **L**. The concept of proof belongs to the syntax of **L**. Furthermore, assume that **M** is a model of **T** (symbolically: $\mathbf{M}^{\mathbf{T}}$). We think about **T** as true in $\mathbf{M}^{\mathbf{T}}$. Accordingly, if $A \in \mathbf{T}$, A is true in **M** (symbolically: $A \models \mathbf{M}^{\mathbf{T}}$. Since we have two objects, $\mathbf{T}^{\vdash}$ (syntactic; provability in **T**) and $\mathbf{T}^{\models}$ (semantic, truth in **T**) and the question arises how to compare both items. The precise answer is given by the completeness theorem:

(**) $A \models \mathbf{M}^{\mathbf{T}}$ iff $\mathbf{T} \vdash A$, for every $\mathbf{M}^{\mathbf{T}}$ and every A (belonging to the language of **T**).

The implication $\mathbf{T} \vdash A \Rightarrow \mathbf{T} \models A$ expresses the property of soundness of inferences (the rules of logic do not produce falsehoods from true assumptions). The dependence $A \models \mathbf{M}^{\mathbf{T}} \Rightarrow \mathbf{T} \vdash A$ states the proper completeness theorem, which holds for all first-order theories. In a sense, (**) justifies the slogan that provability constitutes a criterion of truth, because this theorem establishes the extensional parity of being provable and being true. However, we encounter a problem consisting in the fact that the completeness theorem has no finitary (constructive, etc.) proof. Properties of $\models$ are responsible for that. As I already noticed, syntactical devices, as distinct from semantic ones, are of an effective character.

The concept of completeness as defined above must be distinguished from the syntactic concept of completeness: a theory **T** is syntactically complete iff for any A (expressible in **L** of a given theory **T**), $A \in \mathbf{T}$ or $\neg A \in \mathbf{T}$). As far as the issue concerns mathematics (more specifically, arithmetic of natural numbers), some systems, for instance, the Presburger arithmetic (arithmetic with the operation of addition as the sole operation) or the Skolem arithmetic (arithmetic with the operation of multiplication as the sole operation) are syntactically complete, but some of the most interesting and important mathematical theories, particularly, the Peano arithmetic (**PA**) and its extensions, lack the syntactic completeness. This extremely significant fact was proved by Gödel in his first incompleteness theorem saying that **PA** is incomplete, that is, there exists an arithmetical sentence A, that $A \notin \mathbf{PA}$ and $\neg A \notin \mathbf{PA}$, provided that **PA** is consistent. Adding unprovable (undecidable) arithmetical sentences as new axioms of **PA** does not improve the situation, because the resulting theories remain incomplete. Furthermore, the sentence asserting that **PA** is consistent (con(**PA**)) cannot be proved in **PA** (the Gödel second incompleteness theorem saying that con(**PA**) $\notin$ **PA** and $\neg$con(**PA**) $\notin$ **PA**). The formula 'con(**PA**)' can be eventually proved in a richer system, but this game can be continued *ad infinitum*. How,

---

ferential instrumentarium, the so-called special axioms should be added to (α) – (ε). I assume classical first-order logic.

then, **PA** enjoys semantic completeness, if it is syntactically incomplete? In order to achieve semantic completeness of **PA**, we can assume that all its truths function as the axioms of this theory. Consequently (other version of the first incompleteness theorem), the set of arithmetical truth is not finitely axiomatizable. Consequently, if this set is infinite, we cannot form a finite list of true **PA** axioms. On the other hand, since any proof based on *Cn*–operation (as defined above) is finite, provability serves as a criterion of truth. Clearly, the infinitary parameter introduced into the above description concerns truth, not proof.

However, the situation is still more complicated than the last section suggests. The first Gödel theorem can be also formulated by saying that there are true but not provable arithmetical sentences. Gödel employed the concept of truth in an intuitive manner. In particular, he did not define the concept of truth and relied on its intuitive understanding. On the other hand, the mentioned method of arithmetization envisaged leads to a definition of the concept of proof within **PA**. Consequently, we can define the set of provable numbers (this term was suggested by Tarski 1969), that is, the Gödel numbers of provable sentences. Gödel was aware that the concept of truth behaves in another way (Gödel 1934, p. 363).

> So we see that the class α of numbers of true formulas cannot be expressed by a propositional formula of our system, whereas the class of β of provable formulas can.

This statement can be understood interpreted as a confession that the concept of truth exceeds devices available in **PA**. The issue was completed by Tarski's undefinability theorem asserting (to repeat) that the set of **PA**-truths is not definable in this theory. Thus, we have:

(***) $\mathbf{Prov}^{\mathbf{PA}} \subset \mathbf{Tr}^{\mathbf{PA}}$,

($\mathbf{Prov}^{\mathbf{PA}}$ – the set of provable numbers related to **PA**, but $\mathbf{Tr}^{\mathbf{PA}}$ – the set of true numbers, another label suggested by Tarski; Gödel used a similar terminology). (***) provided a comparison of the concepts of proof and truth (Gödel was conscious of (***), but Tarski proved this fact exactly). Due to incompleteness, the set $\mathbf{Prov}^{\mathbf{T}}$ is not recursive. However, this set is recursively enumerable. Hence, we can (in principle) enumerate the theorems of **PA**, although the set of its non-theorems lacks of this property. But what about the set $\mathbf{Tr}^{\mathbf{PA}}$ as far the issue concerns recursiveness, etc.? This collection does not belong to any level of the arithmetical hierarchy (see Murawski 1991, p. 284). Thus, it is not recursively enumerable. And this observation exactly shows how enormous and deep is the real gap between truth and proof. This observation has additional evidence in the fact that the theoretical realization of syntactic completeness, although it actually overcomes the first incompleteness theorem, the class of $\mathbf{Tr}^{\mathrm{T}}$ remains not definable. Moreover, although both Gödel's incompleteness theorems have effective proofs, Tarski's undefinability theorem does not admit such a proof. Finally, if we add the rule of ω-induction to **PA**, this theory becomes syntactically complete but at the expense of infinite inference rules, contrary to the finiteness condition for *Cn.* On the other hand, the theoretical realization of syntactic completeness, although it actually overcomes the first incompleteness theorem, the class of $\mathbf{Tr}^{\mathrm{T}}$ still remains not definable. Thirdly, although both Gödel's incompleteness theorems have effective proofs, Tarski's undefinability theorem does not admit such a proof. Although metamathematical results, which helpful for a comparison of proof and truth, are rigorously provable, they actually show a difference between syntax and semantics in respect to the amount of being formal of both fields. Since the metatheory related to **STT** is semantic, it includes more informal contents than the metatheory governed by *Cn.*

XVIII
The Paradox of Analycity and Related Issues

Assume that a sentence (I prefer to speak about sentences as bearers of logical properties. By definition, sentences are syntactically well-formed as well as equipped with meaning and thereby intelligible as far as their understanding is taken into account;) A is analytic or is an analytical ($A \in \mathbf{AN}$) under any substantial definition of analyticity (by the adjective 'substantial' I mean here: one of the definition of the concept of analyticity occurring in standard philosophical or/and logical discussions on this topic). Generally speaking, such a definition falls under the following scheme:

(*) $A \in \mathbf{AN}$ if and only $\mathbf{C}(A)$,

where the letter **C** refers to a condition to be satisfied by an analytical. For instance, **C** can state 'is true in virtue of meanings', 'is a tautology', 'is true in all possible worlds', 'it is true on the base of rules of a given language', 'is derivable solely on the basis of logic and definitions', 'its negation is contradictory', etc. (see Woleński 2004d for an extensive survey of various attempts to define the concept of analytic sentence from Kant to the end of the 20th century). However, no particular understanding of analyticity is presupposed in my further discussions, except claims captured by the conditions (1) and (2) formulated in the next section.

In order to make things more explicit, I assume two statements, namely:

(1) A is analytic and true if and only if $\neg A$ is analytic and false; symbolically:
$A \in \mathbf{AN} \wedge A \in \mathbf{VER} \Leftrightarrow A \in \mathbf{AN} \wedge A \in \mathbf{FLS}$ (or $\neg A \in \mathbf{VER}$);
(2) $A \in \mathbf{VER} \Leftrightarrow A$.

Intuitively, (1) and (2) assert the following, presumably non-controversial facts: (a) the concept of analyticity is actually a semantic one (analytic sentences are either true or false; I do not dwell on the problem of analyticity in many-valued semantics); (b) analyticity is closed under the operation of negation, that is, denials of analytic sentences are analytic as well, and (c) the concept of truth obeys the condition regulated by so-called T–equivalences. Although (1) seems fairly trivial, its content has some relevance from the historical point of view. In particular, Kant defined analyticity only for analytic truths (clearly, if 'S is P' is analytic, the content of P cannot be a part of the content of S), but early Wittgenstein and logical empiricists (in the radically syntactic version of their philosophy) did not attribute truth or falsity to analyticals (the issue is more complex, but I skip details). Accordingly, (1) should be explicitly or, at least implicitly, noted in any discussion on analyticals. As far as the issue concerns (2), this assumption corresponds to Tarski's basic requirement concerning the material correctness of the semantic definition of truth.

Consider the sentence

(S) (S) is not analytic.

The sentence (S) asserts (about itself) that it is not analytic. Thus, (S) uses the predicate '(not) analytic' self-referentially.

Suppose that (S) is true. By (2), this assumption entails (S). Thus, (S) is not analytic. On the other hand, ¬(S) is false and, thereby, analytic. By (1), (S) is analytic as well. To sum up, if (S) is true, it is analytic and not analytic. Contradiction! Let us suppose now that (S) is false. Thus, it is analytic. By (1), its negation, that is, the sentence ¬(S), is true as well as an analytical. This implies (by the theorem of sentential calculus: $(A \wedge B) \Leftrightarrow (A \Leftrightarrow B)$) that $\neg(S)$ is analytic iff it is false. Consequently, (S) is analytic if and only if it is true. However, the last assertion entails that (S) is not analytic if and only if it is false. This conclusion is at odds with (1). Contradiction! This completes the argumentation that (S) leads to a paradox, which is analogical to the Liar Antinomy (I use the terms 'antinomy' and 'paradox' interchangeably).

According to the Leśniewski–Tarski diagnosis (see Tarski 1933, Tarski 1944), the Liar paradox has its source in three facts:

(A) Classical (bivalent) logic is applied;
(B) T–equivalences characterize the concept of truth;
(C) Self-referential sentences are admitted.

These points suggest the following ways out:

(A1) Changing logic, for instance, to qualify self-referential sentences as having other logical values than truth or falsity or being truth-value gaps;
(B1) Rejecting T–equivalences as inherently related to the concept of truth;
(C1) Making formal semantic considerations free of self-referential sentences (of course, except reasonings showing their paradoxical consequences).

Tarski decided to adopt (C1) as the most natural solution, because changing logic or rejecting the condition imposed by T-equivalences would cost much more than excluding self-referentiality from semantics. The transfer of semantic concepts to metalanguages connected with given object-languages becomes the outcome of this decision. Thus, if **L** is an object-language and the concept of truth is to be adequately defined for **L**, it must be done in its suitably selected **ML** (since requirements of suitability are not relevant here, I omit this question in the further discussion).

Tarski's solution invoked some criticism. It was argued (see Kripke 1975) that self-referential sentences have a correct grammatical form and, thereby, they are admissible in natural language as well as its formal idealizations. Moreover, for the stratification of the concept of truth (we need several truth-predicates relatively to the hierarchy $\mathbf{L}_0$, $\mathbf{L}_1$ (= $\mathbf{ML}_0$), $\mathbf{L}_2$ (= $\mathbf{ML}_1$), …, $\mathbf{L}_k$(= $\mathbf{ML}_{k-1,}$ …). Finally, not all self-referential sentences are paradoxical. Consider, for instance,

(Q) (Q) does not consist of ten words.

Clearly, (Q) does not cause any trouble. It is evidently false – seven words occur in (Q) – and, its negation, that is, the sentence 'it not true that (Q) consists of ten words', is evidently true. No way of reasoning leads to the conclusion that (Q) is simultaneously true and false. This observation justifies the statement that the sentence (Q) does not lead to a paradox, although it is perfectly self-referential. In fact, safe self-referential utterances occur in colloquial languages much more frequently than their paradoxical cousins.

I will not discuss all the objections against Tarski's solution. Let me observe, at first, that we do not need to qualify self-referential sentences as meaningless. Even if some of Tarski's remarks suggest that the liar sentence and similar utterances are incorrectly formed from the syntactic point of view, this proposal seems to be artificial. In fact, if we infer something from the liar formula or (S), it is difficult to maintain that our fundamental premise has no meaning. Polish logicians, including Tarski himself (although he was not always transparent in his remarks on (in)correctness of the liar sentence), considered self-referential sentences rather as items which could not be values of sentential variables. Not because they are grammatically defective, but for their creative role in generating semantic antinomies. Using perhaps a somewhat exaggerated comparison, one could say that, similarly, we should not divide by 0 or that the absolute vacuum is physically impossible and cannot be measured in any reasonable system of magnitudes. However, we perfectly understand what to divide by 0 means or how to measure the absolute vacuum. Mathematical or physical impossibilities do not necessarily imply logical meaninglessness.

My main focus in the coming remarks is to investigate, or at least to indicate, the role of T–equivalences in producing semantic troubles in which such equivalences are involved. If we look at the steps leading to the Liar Antinomy or the analyticity paradox, we immediately observe that T–equivalences play an absolutely crucial role in this business. Roughly speaking, T–equivalences compress the left-hand part of the formula $TA \Leftrightarrow A$ to its right-hand part. The analyticity paradox and the Liar Antinomy cannot be derived without such a compression. On the other hand, the sentence (Q) has no such connection with the related T–equivalence, that is,

T(Q) if and only if Q, which produces a semantic antinomy in the case of the liar sentence or (S). It seems (it is an empirical claim) that T–equivalences lead to semantic troubles only in the case that they compress semantic concepts. This assertion is very nicely supported by the derivation of the Liar Paradox in formalized Peano arithmetic, when we use the fix point (or diagonal) lemma. Let the symbol **PA** refer to formalized (first-order) Peano arithmetic (see Smullyan 1992, pp. 102–105). The diagonal lemma says that **PA** proves a formula $F(A^*) \Leftrightarrow A$, where A^* is the Gödel number of the formula A and F is an arbitrary property expressible in arithmetic. In particular, **PA** proves T-equivalences, if F is interpreted as 'is true'. Since the diagonalization function $f(A^*, A)$ is (strongly) definable in **PA,** the fix point lemma guarantees that T–equivalences for self-referential sentences are available as theorems of **PA**. However, since the same procedure leads to their negations, we immediately obtain the Liar Paradox. Thus, in order to avoid the paradox we should either exclude some T–equivalences or admit that the truth-predicate is only partially defined. If we exclude self-referentiality, the predicate T becomes coherently definable, but we must go to **ML** in order to provide a consistent definition. This conclusion follows from Tarski's undefinability theorem: if a theory **T** is sufficiently enough to cover **PA**, its truth-predicate is not definable in **T**. Note that the formal construction using the diagonal lemma can be fairly considered as at least partially parallel to Tarski's analysis of the Liar paradox in natural language. This observation supports the view that formal constructions do not depart very dramatically from ordinary usages.

Coming back to analyticity, the paradox concerning (S) combines analyticity and truth via T–equivalence T(S) if and only if (S). In fact, T–sentences compress analyticity to truth or falsity. Since we assume that analytic sentences have definite logical values, that is, they are true or false, it is not strange that both compressions produce antinomies. What is a solution? Since the diagnosis of the sources of the trouble with (S) closely follows the points (A) – (C), possible ways out are indicated by the recommendations (A1 – (C1). Consequently, if we intend to keep T–equivalences as determining the use of the predicate 'is true' and classical logic as our logical environment, the simplest proposal consists in claiming that in order to define analyticity for **L**, we must use resources available in **ML**. This suggestion not only follows directly from the assumption (1), but is also supported by formal constructions related to the very nature of semantic concepts. If classical logical values, that is, truth or falsity, are attributed to analyticals, no other way looks reasonable, unless we resign from considering them as genuine semantic items. Anyway, the real strength of the Tarskian (model-theoretic) semantics is sufficient enough to produce a general uniform frame for synthetic as well analytic sentences.

XIX
Some Liar-Like Paradoxes

The classical Liar Antinomy or Paradox (**LA**, for brevity) runs as follows (it is a slightly modified version proposed in Poland by Jan Łukasiewicz and employed by Alfred Tarski):

(1) The sentence denoted by (1) is false;
(2) (1) ⇔ (1) is true;
(3) The sentence (1) is false ⇔ The sentence (1) is true;
Contradiction!

Remark: How to understand the premise (2)? Some commentators say that it is a nonsense, because it assumes that the equality (1) = (1) is true, but this equation is plainly false, if identity is understood in its official logical meaning. Imagine, however, that we correlate some objects with numbers (numerals), for example, saying that Lionel Messi has the number 10 as a soccer player in FC Barcelona or the Argentinian national team. Consequently, we are entitled to say 10 = Lionel Messi in such and such context. The same can be said in the case of **LA**. What is important, is that the equivalence (b) is justifies by this convention.

Stanisław Leśniewski, followed by Tarski, offered a diagnosis of Leśniewski–Tarski. According to them, we can identify three sources of **LA**:

(i) Self-referentiality of the predicate 'is false';
(ii) T-scheme, that is, the formula *A* is true ⇔ *A*;
(iii) Classical logic.

This diagnosis opens related possibilities concerning ways of avoiding **LA**. The first consists in rejecting self-reference of semantic predicates, the second – banishes T-equivalences, and the third goes by changing logic. According to Tarski, T-scheme is too intuitive to be rejected and classical logic should be protected as the best. Consequently, rejecting self-reference of semantic predicates leads to the least theoretical costs. A natural outcome of the first strategy consists in the language (**L**) / metalanguage (**ML**) distinction. An important lesson stems from the above analysis: nothing is free of charge, also in logic.

One can ask whether elimination of self-reference concerns only the predicate 'is false'. What about the predicate 'is true'? We can describe both of them as aletheiological words. It was proved that the formula:

(*) (1) ⇔ (1)

is true and does not lead to any paradox. However, if we change logic into so-called dual logic (or another in which the logical value of falsehood is distinguished; *A* is a dual tautology if and only if it is false for any valuation), (*) leads to the dual **LA** or the Truth-teller paradox (see Woleński 1995 for details). Thus, both aletheiological predicates produce semantic troubles. Changing logic does not block the strengthened **LA**. It is captured by:

(a) the sentence (1) is not true (false or other);
(b) (1) ⇔ (1) is true;
(c) (1) is true ⇔ (1) is not true;
Contradiction!

If (1) is *n* (other than false or true), we get (1) is *n* ⇔ *n* is true, which also cannot be accepted.

A very important observation is this. T-scheme plays the fundamental role in deriving **LA** (in all of its forms). On the other hand, there are cases which are not paradoxical. Consider an anthropomorphic sentence saying 'I am provable' or 'I consist of 3 words'. Clearly, we cannot formulate relevant T-schemes. The relation of truth and provability is captured by Löb's theorem, that is, the formula 'if *A* is provable, *A* is true'. The reverse implication does not hold, due to the Tarski undefinability theorem. Turner (see Turner 1990, Turner 1990a) observed that the only part of T-equivalence, that is, the formula '*A* ⇒ *A* is true', is responsible for **LA**. Contrary to important uses of the relation between truth and provability in metamathematics, any attempt to form a T-sentence for 'I consists of 3 words' (falling under the scheme '*A* is true if and only if *A*') leads to an obvious absurdity, because omitting 'consists of 3 words' converts the entire sentence into a nominal expression if we decide to treat *A* as an abbreviation for '*A*' or reduces this sentence to itself.

Are 'is true' and 'is false' the only predicates, which produce paradoxes similar to **LA**? The answer is negative and I will show that some other predicates share the fate of it. Historically speaking, the self-referential use of the 'is meaningful' was suspected to be troublesome. This problem arose as connected with the question "Is the verifiability principle as proposed by logical empiricism, meaningful or nor?" (see Ingarden 1936; Ewing 1937, Woleński 2010).

Logical empiricists argued that we should accept:

(PV) A sentence *A* is meaningful if and only if *A* verifiable or tautological.

Ingarden and Ewing asked whether (PV) itself is verifiable or tautological; I neglect the difference between both philosophers in their criticism in question. According to Ingarden and Ewing, both alternatives are untenable, because it is difficult to imagine how (PV) could be verified by empirical data, but, on the other hand, if this principle is a tautology, it is devoid of meaning. Consequently, (PV) should be considered as devoid of meaning by own criteria or, as Ingarden pointed out, the principle in question smuggles a hidden meaning to metalogic and thereby remains unclear.

Carnap (see Carnap 1936) briefly commented on Ingarden's criticism at the Prague Congress in 1934 and, coherently with his treatment of logic as syntax, argued that metalogical sentences belong to pure syntax and, thereby, they do not require empirical verification. Consequently, metalogical sentences are tautological just by definition and it means that their validity can be established by formal combinatorial procedures. Moreover, Carnap stressed that one of the most crucial differences between logical empiricism and phenomenology consists in admitting *Wesenschau* by the latter school and rejecting this kind of cognition by the former philosophical group. Although Ingarden could explain the hidden meaning of metalogical sentences by recurring to phenomenological intuition, but nothing essentially depends in Ingarden's criticism on a specific basis of the hidden meaning. It is sufficient to observe that metalogical sentences are meaningful by virtue of other criteria than those stated by (PV). Both parties can also agree that empirical verification has nothing special to do in metalogic. In fact, Carnap and other logical empiricists accepted this position and looked for a solution via the concept of analyticity. Pap in his comments on Ewing 1937 (see Pap 1957) proposed a way out, consisting in considering (PV) as a result of an analysis of the concept of meaning. Pap remarked that this view overcomes the objection that (PV) says nothing at all, being a logical tautology. Actually, this proposal weakens Carnap's original idea that metalogical assertions.

In order to formulate the Liar-like paradox for 'is meaningful', we need the following additional principles:

(4) A is meaningful $\Leftrightarrow$ $\neg A$ is meaningful;
(5) A is meaningful if and only if A is true or false.

The principle (4) says that meaningfulness is invariant under negation. Its justification can appeal to the idea that purely logical operations do not influence the logical status of their arguments. Consequently, if A is a logical sentence (a tautology or a contradiction), its denial is a tautology or a contradiction as well, and, if A is factual (neither a tautology nor a contradiction), its negation is factual as well. Yet, negation changes the logical value and (5) exhibits the form of meaningfulness related to classical logic.

Having the statements (4) and (5), we proceed in the following way. Firstly, we search for a sentence with the self-referential employment of the predicate 'is not meaningful'. This step is given by:

(6) (6) is not meaningful (assumption).
(7) $\neg$(6) is not meaningful (by (4)).

On the other hand, (7) says that (6) is meaningful. Consequently, (6) is either true or false by (5). Assume that (6) is true. It gives:

(8) (6) is true $\Leftrightarrow$ (6) (an instance of T-scheme).

By the detachment, we obtain (6), but it produces a contradiction, because (6) appears as true and not meaningful.

Assume that (6) is false. Hence, (6) is meaningful, but $\neg$(6) is meaningful and true. If so, we have:

(9) $\neg$(6) $\Leftrightarrow$ (6) is meaningful.

As a consequence we obtain

(10) (6) $\Leftrightarrow$ (6) is not meaningful.

However, this step convert us to the former case, because it implies that (6) is meaningful if and only if it is true.

If we say that being not meaningful covers a third value, we obtain the strengthened meaningfulness paradox. Ewing (see Ewing 1937) observed another difficulty. Consider (PV) and ask what happens if we check its truth or falsity without taking into account sentences like (6) that is asserting its own meaninglessness. Clearly, (PV) should be meaningful under its truth as well as under its falsity. However, the latter assumption leads to a strange consequence that (PV) is false and meaningful, but in such a case it defeats itself. It is not a logical paradox because not all eventualities lead to formal inconsistency (more specifically, it is the case of working under the assumption that (PV) is true), but a plain pragmatic or epistemological oddity. Another strange consequence follows from inspecting (7). The negation of (6) means that (6) is meaningful. If so, asserting (7) commits us to the view that (6) is meaningful, provided that it is not.

Another example of a Liar-like paradox concerns the concept rationality. As an example, I consider what can be called Cartesian rationality defined by:

(11) (a) *A* is rational iff its truth is dictated the infallible method;
(b) *A* cannot be rational and false.

We can form the self-referential sentence:

(12) (12) is not rational (its truth is not dictated by the infallible method),

which leads to:

(13) (12) is true ⇔ (12) is not rational.

By the propositional calculus, (12) either true by the infallible method and not rational or false and rational. However, both these possibilities are contradictory. Presumably, one could think that other definitions of rationality suffer from the paradox. Unfortunately, this is not the case. Let us adopt:

(14) *A* is rational ⇔ *C*,

where the letter *C* refers to a condition of rationality. Thus, we have:

(15) *C* is rational ⇔ *C*;
(16) C is rational ⇔ *C* is true.

Now, if we consider:

(17) *C* does not satisfy the condition *C*,

we obtain a version of a Liar-like paradox. The situation can be easily generalized for such predicates as 'it is analytic', 'it is confirmed', 'it is scientific', etc. A hypothesis can be stated that almost every interesting epistemological predicate falling under the scheme:

(18) *A* is *X* ⇔ *C* (⇔ *C* is true),

leads to a Liar-like paradox. (18) confirms that the role of T-scheme in Liar-like paradoxes is crucial. Perhaps, the predicate 'is dubious' appears as particularly interesting, because it is related to the issue of skepticism.

A general situation looks as follows. If a given principle P establishes the meaning of a predicate *P* referring to properties of sentences such that T-scheme is applicable, we can expect that the predicate in question can generate a Liar-like paradox. In order to prove that *P* does not produce a paradox, one must prove that 'is *P*' and 'is true' do not coincide. It is the case of 'is provable'. Taking a lesson from metamathematics, *P* is not paradoxical, if there is a procedure, which allows to embed this predicate into the object-language via a procedure analogical to arithmetization. Unfortunately, it is unclear how typical epistemological predicates could be converted into the object language.

However, it does not mean that philosophers must resign from *P*. Generalizing the case of using 'is true' in order to avoid the Liar-paradox, we can decide that *P* is formulated in **ML** and apply it to items formulated in L. Yet, this simple prescription leads to far-reaching metaphilosophical consequences. Take, for example, the problem of how science and philosophy are mutually related. The claim that philosophy (in our case, epistemology) is one of sciences, requires an assumption that (PV) and similar principles are embeddable into L as it is used in science. As we know, this assumption is not obvious. I will leave this fundamental problem without further considerations. Still, one remark is in order. Pap's view that (PV) offers an analysis of the concept of meaning leads, as I already remarked, to the question "What about the status of this principle and similar ones?" A preliminary and very tentative answer is that such analytical products are analytical statements, not reducible to tautologies of **ML**. Otherwise speaking, if *C* is a criterion of analyticity in **L**, other (in fact, extended) understanding of 'is analytic' functions in **ML** (see Woleński 2004d for some proposals in this respect).

XX
Appendix
Nothingness, Philosophy and Keret's House

The Israeli writer Yehiel De-Nur, also known by his pen name Ka-Tzetnik, was an extermination camp survivor who testified against Adolf Eichmann at the 1961 trial in Jerusalem. He said there, among other things:

> I was there for about two years. Time there was not like it is here on earth. Every fraction of a minute there passed on a different scale of time. And the inhabitants of this planet had no names, they had no parents nor did they have children. There they did not dress in the way we dress here; they were not born there and they did not give birth; they breathed according to different laws of nature; they did not live – nor did they die – according to the laws of this world. Their name was the number...[49]

One could hardly imagine a more moving testimony of that topsy-turvy world, another world, or world of death, where virtually everything was different than in the normal reality, the world of life. One of the German doctors working at Auschwitz told his colleague before an *Aktion* in the women's barracks that they would find themselves in the *anus mundi* – 'anus of the world'. In fact, everything there was organized so that the 'inhabitants of this planet' would never forget that their only duty was to die (as put by one of the Treblinka commanders).

It immediately prompts one to consider this world of death as an annihilating reality. In his statement, Ka-Tzetnik referred to Auschwitz rather than the Warsaw Ghetto, on the former site of which the Keret House is located today. What was officially known as the 'Jewish Housing District' was not an extermination camp. But it was a sort of entrance to it. There is no doubt that conditions in the Ghetto were different than those in Auschwitz. Still, the atrocities that took place in the former mean that it was also a world far detached from the normal reality. This is clearly attested to by Yael Hersonski's *A Film Unfinished*, which uses footage filmed by German cameramen in the Ghetto. The different episodes are commented on by five survivors of the Warsaw Ghetto. The viewer gets an irresistible sense that these people are horribly afraid of what they had to experience and what is now presented to them. Some scenes, such as that of a solemn

---

49 The Trial of Adolf Eichmann, Record of the Proceedings in the District Court of Jerusalem, Session 68, http://www.nizkor.org/hweb/people/e/eichmann-adolf/transcripts/Sessions/Session-068-01.html (accessed: 25 January 2014).

funeral ceremony, were staged by the Germans to show that life in the Ghetto was not far from normal. One of the survivors comments on the scene by saying that no one received a proper burial in the Ghetto. People died in large numbers and the streets were filled with abandoned corpses, which would then be placed in mass graves. The Warsaw Ghetto was to be annihilated. 'The former Jewish quarter of Warsaw is no longer in existence', SS commander Jürgen Stroop wrote in an internal report after the Ghetto Uprising had been crushed.

The Keret House has been built in a narrow clearing between the buildings at 22 Chłodna Street (a postwar apartment block) and 74 Żelazna Street (a prewar tenement) according to a design of the architect Jakub Szczęsny (Grupa Projektowa Centrala Studio). It is a temporary residency space for artists, and the residents are selected via a competition procedure. There are no restrictions regarding the subject of their prospective project and the artists are free to pursue whatever artistic scenarios they come up with. These scenarios are not expected to refer in any way to the Holocaust. Still, various associations with the past are understandable. After all, as has been said, the Keret House is located on the site of the former Ghetto, near the famous wooden footbridge that once connected the Small Ghetto (in which mixed, that is, Jewish/Non-Jewsih families lived) with the Large Ghetto (for "purely" Jewish people; living conditions in this area were worse than in the Small Ghetto), known as the Bridge of Sighs, for those crossing it were able to see a glimpse of what in comparison with their own existence was the life-almost-as-usual going on underneath.

The Keret House's first resident was Etgar Keret (hence the name), an Israeli writer whose mother was a survivor of the Warsaw Ghetto and whose father survived the war hiding near Warsaw. Keret himself does not directly address the Holocaust in his works and this could suggest that no interpretation informed by the past or present of the building at the corner of Chłodna and Żelazna Streets is needed, or even recommended. And yet the situational context of the location of the Keret House, the common knowledge that the 'Ghetto used to be here' and finally the person of the place's first resident – all invite a broader reflection. Let me admit at this point that I have strong doubts whether this text, very much philosophical in nature, has anything to do with the Keret House installation. I hope that the final parts of this essay will show that my doubts can be overcome.

When I was asked to write an essay about the Keret House, using the notion of nothingness seemed inevitable. As we read in the seminal *Encyclopaedia of Philosophy*,

> Nothing is an awe-inspiring yet essentially undigested concept, highly esteemed by writers of a mystical and existentialist tendency, but by most others regarded with anxi-

ety, nausea, or panic. Nobody seems to know how to deal with it (he would, of course), and plain persons generally are reported to have little difficulty in seeing, hearing and doing nothing. Philosophers, however, have never felt easy on the matter.[50]

The history of the notion of nothingness confirms this. Since antiquity up to the present time it has been the subject of all kinds of reflections. Their selection below does not pretend to be complete, nor does it contain typically academic erudite material.

Parmenides was probably the first to pose the problem of nothingness in general. As he famously stated (I follow Barnes' translation; Barnes 1979, p. 157),

> Come here, I will tell you (and you must spread the story when you have heard it) what are the only roads of inquiry for thinking of:
> One, both that it is and that it not for no being, is the path of Persuasion (for Truth accompanies it);
> The other, both that it is not that it is necessary for it not to be.

This rendering follows colloquial Greek. A more sophisticated version, popular among many philosophers, renders Parmenides' statements as saying that Being is (eventually, must be), but Non-Being does not exist (eventually, is impossible). Parmenides also asserts that being is eternal and stable. This is interpreted as a principle which in Latin was expressed as *ex nihilo nihil fit*, that is, 'nothing comes from nothing' (or, more strongly, 'nothing can come from nothing'). From Parmenides' notion also arises the principle of *natura horror vacui* (nature abhors a vacuum), accepted by Plato, Aristotle and the subsequent naturalists.

The question is whether this principle is universal. Even if we assume that it holds true in the natural sciences, and especially in physics (as evidenced by statements that absolute vacuum does not exist but there exist its various experimental or empirical approximations, and by the acknowledgment of those physical laws that deal with frictionless movement or free fall as theoretical idealisations fulfilled in appropriate mathematical models), it does not have to apply to philosophy. And indeed, the notion of nothingness entered philosophy through various ways. One led from theology and its belief in *creatio ex nihilo* (creation out of nothing). While the dominant tradition in Judaism holds that this is how the Almighty created the world, certain passages in the Hebrew Bible suggest that the act of creation started not with nothing but with chaos. Christianity accepts unreservedly (it is even a Catholic dogma) that God called the world into existence out of nothing. This view stood in contradiction to Greek

50 Kerferd 1967, p. 324. For simplicity, 'nothing' and 'nothingness' are considered as practically synonymous in this paper.

philosophy on the strength of the *ex nihilo nihil fit* principle. Paul of Tarsus met with Greek philosophers (Stoics and Epicureans) in Athens during his second mission journey dated for CE 50–52. According to The Acts of the Apostles (17:24), he delivered a sermon in the Areopagus, where he said, among other things, that 'God that made the world and all things therein'. Although these words do not imply directly that a *creatio ex nihilo* had been the case, the phrase 'the world and all things therein' may be construed in the spirit of the later dogma of a creation out of nothing. The 'men of Athens', as Paul addressed them, were probably not convinced. Although their words, 'We will hear thee again of this matter' and even mockery pertained to his preaching about resurrection, we may suppose that they reacted similarly to statements contravening the *ex nihilo nihil fit* and *natura horror vacui* principles.

The notions of being and nothingness (viz., non-being) gave rise to various conceptual and even theological problems. Seeking to address them was the mediaeval theory of the transcendentals, that is, of 'most common notions', such as (allowing for variations between the different scholastic schools) 'being', 'true', 'good', 'beautiful', 'one' or 'thing'. According to the most popular conception, originated by Thomas Aquinas, if *t* and *t'* are transcendentals, then their scopes overlap, so, for example, if something is a being, it is also true, and if something is true, it is also a being (*ens et verum convertuntur* – 'being and true are convertible'). The notion of being is singled out here insofar as the other transcendentals express some aspects of being-ness, such as that every being is good or beautiful. Transcendental notions possess a certain peculiarity: they do not refer to genera, that is, to sets denoted by general names, such as 'man', 'animal' or 'house'. In the light of traditional ontology, deriving from Aristotle, genera form a hierarchy, the levels (viz., the different genera) of which are defined using the method of the nearest genus and generic difference; for example, we define man as an animal (the nearest, or superior, genus) that is rational (generic difference). The nearest genus can be divided dichotomously: animals divide into humans and non-humans. However, being is the most general notion. This means that it has no superior genus. Consequently, there is nothing that could be divided into being and non-being. Technically speaking, the complement of the class of beings is empty. There is, therefore, no non-being, and thus no nothingness. This part of Parmenides' theory was preserved in the theory of the transcendentals. The latter, however, had to confront the problem of *creatio ex nihilo*. If we accept that there exist bad and ugly things, then we must assume that they have been created by God, since He has called everything into being. But this is out of keeping with His goodness. The solution is to accept that if being, goodness and beauty are co-extensional (identical in scope) with being, then their negations – evil and ugliness – are nothingness. It is logically impossible that God could not have created nothingness. So how to explain the presence of things evil and ugly? The scholastic construed the negation of the transcendentals as a lack (privatively). Non-being (nothingness) is a lack of being, evil is a lack of goodness and ugliness is a lack of beauty. Consequently, nothingness, evil and ugliness are not God's creations but human faults.

Another channel through which the notion of nothingness entered philosophy was a question posed by Leibniz: 'Why is there something rather than nothing?' (Some philosophers, such as Martin Heidegger, regard this as the single most important philosophical problem bar none). Leibniz did not claim that nothingness exists. The very posing of the question seems to imply that it could be so that there would exist a perfectly empty world, that is, a world without being. Leibniz, as we know, postulated the category of 'possible worlds', construing possibility as a logical non-contradiction. From this ensues that an empty world (non-being, nothingness) is logically coherent, although the universe we live in is, of course, non-empty (there is something in it rather than nothing). According to Leibniz, nothing is simpler than something, and if so, we need to ask why there exists a more complex world rather than an empty one. This argument is problematic, though. Negation (the root 'no') is contained in the very term 'nothing' (nothingness), but not necessarily so in 'something' (being). This suggests that being is a simpler category than nothingness. The fact, however, that being has a more or less complex structure is irrelevant here. If we assume that from the logical point of view nothingness is more complex than being, then we should reverse Leibniz's question and ask: 'Why should there be nothing when there is something?'

The idea of nothingness suggested itself also in the context of change and temporality in the world. Jean-Paul Sartre wrote in his famous *Being and Nothingess* (Sartre 2014, p. 107):

> Temporality is evidently an organized structure. The three so-called „elements" of time, past, present, and future, should not be considered as a collection of „givens" for us to sum up – for example, as an infinite series of „nows" in which some are not yet and others are no longer – but rather as the structured moments of an original synthesis. Otherwise we will immediately meet with this paradox: the past is no longer; the future is not yet; as for the instantaneous present, everyone knows that this does not exist at all but is the limit of an infinite division, like a point without dimension.

Speaking colloquially, the past is gone, the future has not yet happened, and the present may be likened to a vanishing point, to use the title of Richard C.

Sarafian's 1971 movie. A point that both is and is not because it keeps eluding us. When I say, 'I am writing this sentence now', as soon as I place the closing quotation mark, I no longer refer to the present but to the past.

Since the times of Heraclitus, the phenomenon of change has presented philosophers with a dilemma. Heraclitus said that it was impossible to enter the same river twice. We may thus posit that change permanently annihilates that which it changes. Parmenides, who disagreed with Heraclitus, concluded from this that the existence of being consists in its immutability. Hegel, in turn, a continuator of the idea that everything changes (*panta rhei*), argued that being and nothingness are identical. Leaving these philosophical complexities aside, let me note that Sartre's reasoning above can be generalized. The paradox of a past that no longer is, a future that not yet exists, and the present as a vanishing point does not apply to space. While I cannot return to the past, I can find myself again in a place of the same spatial coordinates. However, if we take space-time as a point of reference, the past space-time, the present (current) space-time, and the future space-time are fundamentally different.

To derive nothingness from temporality (or, alternatively, spatio-temporality) and change is hardly convincing. The paradox noted by Sartre is in itself impressive, but predicated on an insufficiently thorough analysis of temporal relations. It is immediately evident that whereas the temporality of the past and the future are construed in terms of segments here, as stretches of variable end points (leaving aside the further complications resulting from whether time is construed absolutely or relativistically), the present is treated as a non-dimensional (sizeless) point. There is no reason, however, for what is present to be a point, for it may be construed as being contained within a segment. Although the bounds of the present are blurry, locating them is a matter of convention. At the root of Sartre's paradox is the same problem that preoccupied Zeno of Elea when he formulated his famous aporias (paradoxes) of motion: the inability to understand that an infinite set of points is contained within a segment. Nothingness needs not to be invoked in connection with change. It is true that when objects change over time, they lose some characteristics and acquire others. It is also true that some beings lose their existence. You may call this nihilation, but this is just using a word, not pointing to a new problem. This does not mean, of course, that there are no conceptual difficulties concerning the temporality and mutability of existing beings. There are many conceptions of the relations between parts of objects and their wholes, one of them positing that while parts change, whole objects do not. Another classical problem is that of the identicalness of objects passing through various temporal stages. Let me stress once again, however, that an analysis of these issues does not have to refer to the notion of nothingness as a radical negation of being.

Heidegger and Sartre became the leading protagonists of contemporary philosophy of nothingness. More precisely, in the context of the philosophical thought to date, they developed notions of nothingness on an unprecedented scale. Their theories, usually denoted by the umbrella term of existentialism, consisted in ontological analyses in the classic sense, that is, being as being, but with elements of philosophical anthropology. The issue of nothingness is, however, significant only for being-in-itself (Heidegger's *Dasein*), that is, one that consciously experiences its being, the being of being. It needs to be added that being-in-itself cannot be otherwise, that is, without the consciousness that its being is experienced by it. The consciousness of its own finiteness and of the mechanism nihilating *Dasein* is the most constitutive attribute of being-in-itself. All beings divide into those whose essence precedes their existence and those who existence precedes their essence. The latter are beings-for-themselves, opposed in this regard to beings-in-themselves. Their centrality in being as a whole is that they – and only they – are able to understand their distinctness vis-à-vis beings-in-themselves. Contrary to common opinion, the name 'existentialism' is meant to stress not the particularly humane issue of the relation with value or freedom but, above all, the fact that existence determines essence, not the other way round.

Perhaps, the most synthetic explication of the existentialist notion of nothingness can be found in Heidegger's *What Is Metaphysics?*. Elaborating on why there is anything at all rather than nothing whatsoever, Heidegger writes (Heidegger 2014, pp. 35–50):

> Only be-ing is supposed to be studied, and besides that – nothing; only be-ing, and more than that – nothing; solely be-ing, and beyond that – nothing.
> How do things stand with this no-thing? Is it an accident that we speak quite automatically in this way? Is it then only a manner of speaking – and nothing more?
> But why do we trouble ourselves with this no-thing? In fact, no-thing is indeed turned away by science and given up [on] as the null and void. But if we give up on no-thing in such a way, do we not indeed accept it? But can we talk about an acceptance if we accept nothing? . . .
> . . . a question has already presented itself. The question merely needs to be articulated. How do things stand with no-thing?. . .
> . . . What is no-thing? Our first approach to this question already shows us something unusual about it. From the outset in asking this question we posit no-thing as something that 'is' such and such, as be-ing. But plainly it has in fact been distinguished from just that. The question about no-thing – what and how it, no-thing is – turns what is being questioned into its opposite. The question robs itself of its own object.

> Accordingly, every answer to this question is impossible from the outset. For it necessarily starts out in the form: no-thing 'is' this or that. Question and answer alike are themselves just as nonsensical with respect to no-thing. .
> But can the rule of 'logic' be challenged? For no-thing is the negation of the generality of be-ing, simply not be-ing. . . . Is there no-thing only because there is the not, i.e., negation? Or is the other way round? Is there negation and the not only because there is no-thing? . . . We maintain that no-thing is more original than the not and negation. . . .
> . . . If no thing is to be questioned in the way questioning works, then it must itself be given in advance. We must be able to encounter it.
> How do we go after no-thing? How do we find no-thing? . . .
> . . . we know no-thing . . .
> Dread reveals no-thing. . . .
> . . . that about and of which we were in dread was 'really' nothing at all. Indeed, no-thing itself as such was there. . . .
> How do things stand with no-thing? . . .
> . . . No-thing nihilates of its own. . . .
> With that the answer to the question about no-thing is found. No-thing is neither an object nor, above all, be-ing. No-thing comes neither in and of itself nor along with be-ing, upon which it depends all the same. No-thing is the potential for a manifestness of be-ing as some such thing for human existence. No-thing does not primarily provide the antithesis of be-ing, but is originally of the very essence. The nihilation of no-thing happens in the *be*[-ing] of be-ing. Dread is what means that nothingness is given originally. Dread is a mood of Dasein as such. It cannot be identified with a fear of anything, although the fear of death may be considered as its (viz., dread's) exemplification. Rather, dread is an ontological leftover in the being of being-for-itself generated by the nihilation by nothingness. Heidegger insists on speaking of nihilation rather than of annihilation, for being-for-itself approaches nothingness in acts of metaphysical dread.

My view of Heidegger's and Sartre's metaphysics of nothingness is utterly negative. I believe that construing nothingness as a negation of being as a whole and as being originally given in metaphysical dread stems from an elementary incomprehension of simple logical facts. In order to demonstrate this, I will consider Sartre's three characteristics of being-in-itself (I leave aside to what extent they apply to being-for-itself), namely: (a) being is; (b) being is in-itself; (c) being is what it is. Let us now try to replace 'being' with 'nothingness'. We receive: (a1) nothingness is; (b1) nothingness is in-itself; (c1) nothingness is what it is. In the light of Heidegger's explanations, (a1) is valid; (b1) rather incomprehensible (it is not clear whether the distinction between nothingness-in-itself and nothingness-for-itself is valid at all); and (c1) is what worried Heidegger, for it suggests that nothingness is something, even though we assume that it is nothing as a negation of being. Heidegger stated further that nothingness is what nihilates, and, as such, is something. We may assume, of course, that the word 'is' has acquired a special meaning in (c1), allowing Heidegger to overstep logic. This is hardly a major success, though, for the phrase 'special meaning' explains little unless we define that meaning. Nor is it so that the question 'What is nothingness?' cancels itself out, for we may reply, traditionally, that nothingness is nothing. Such a reply has a completely definite logical form, namely, it falls under the following assertion:

(*) There does not exist such an x that x is nothing,

which is a simple consequence of the traditional ontological thesis that every being is something. Heidegger fell into a well-known trap: to argue that *a* does not exist (is nothing), we must first acknowledge that *a* somehow is. In other words, Heidegger erroneously assumed that using the noun 'nothingness' in (a1) and (c1) automatically reifies nothingness (one should observe the quotation marks and the absence thereof). Numerous critics have pointed out that when saying 'Nothing nihilates' (*Das Nichts nichtet*), he failed to recognize the logical form of this statement, mistaking the grammatical subject for a logical one. If we use standard logic, *das Nichts*, as an individual (proper) name, has to be excluded, for every such name necessarily means something, and thus is not empty. The sentence in question, becomes false when construed as:

(**) There exists such x that x nihilates.

If Heidegger were to follow through on his own logic, he should have said that nothingness does not nihilate, expressing a true, if banal, statement. If we allow empty proper names – which are the domain of free logic – the sentence 'Nothing nihilates' becomes an ordinary tautology, for we are forced to define nothingness as something that nihilates, or use (c1) and recognise the analysed statement as an abbreviation of 'Nothingness is that which nihilates, that is, that which nihilates'. In any case, Heidegger's conviction that he overcame logic by discussing the nihilation of nothingness is premature.

In fact, the problems Heidegger encounters when trying to explicate statements about nothingness are nothing new. The discourse of that which is not at all has always been a sensitive matter. If we ask about the meaning of *creatio ex nihilo*, the answer is not simple at all. The common meaning of the verb 'create' is that if x creates something, it does so out of something. Let us assume – for example, on the strength of the appropriate dogma – that God created the world out of nothing, that is, out of nothingness. This means that a special meaning has been given to the term 'create'; it is not clear what this meaning is, but it is certainly different from the colloquial one, encapsulated in, for example, the *ex nihilo nihil fit* principle. If the Athenian intellectuals frowned upon Paul's words about the creation of the world (out of nothing, it could be implied), then they surely had good reasons for that. Problems arise also on the grounds of the aforementioned theory of the transcendentals, which aimed, among other things, to explain why God cannot be regarded as the creator of evil and ugliness. Still, it is hard, at least in the common sense, to accept that evil and ugliness do not exist, if we encounter them all the time. The idea that being is good and beautiful is hardly defensible without making various assumptions, and the introduction of a private notion of nothingness changes little from the philosophical point of view.

Returning to Heidegger, he doubtless took advantage of the lexical possibilities of the German language, where every noun can be transformed into a verb. This means that the sentence *Das Nichts nichtet* appears perfectly natural. In other languages, things are more complicated and in English the sentence has been rendered as 'nothing nothings' or 'nothing noths', and in Polish as *nicość nicuje*, *nicość niczeje*, *nicość nicestwi* or *nicość unicestwia*. All these phrases sound artificial, as does the version I use, 'nothing nihilates'. If I were to explain the failure of Heidegger- and Sartre-style metaphysics of nothingness, I would say that both of them reified Nothing (the capital letter indicates that this means *das Nichts*) by resorting to a rather banal grammatical procedure. Reified nothingness allows sentences of the type '*a* is *b*', for example, 'nothing nothings', and even makes them true.

In Parmenides or Leibniz, nothingness is absolute. It is non-being, that is, a negation of being. The sentence 'nothingness is nothing' is the only sensible statement that can be made about nothingness. It is (see above) a negative statement. Nothingness is neither colourful nor bland, neither great nor small, neither finite nor infinite, etc. – but these are banalities resulting from the fact that nothingness is nothing. As an utter lack of being, nothingness is a completely non-empirical notion: it can neither be touched nor calculated, it does not have a structure, nor is it gradable, and so it cannot be illustrated or approximated by phenomena of this world. If we say, for example, that nothingness is a boundary approximated by a sequence of empirical systems of an ever smaller number of constituents, this will be but a metaphor. Perhaps, the temptation to search for nothingness originated from the earlier notions of time and space, construed as infinite containers for the empirically accessible. It is necessary, however, to seriously consider the lesson stemming from the idea of an expanding universe. It does expand, but together with space. From this ensues that the question about what is beyond the boundaries of the universe does not make sense. This stands in open contradiction to the common notion that if there are boundaries, there has to be something beyond them. Yet, if we adopt the model of a closed but infinite universe (which is not the only model coherent with physics), talking about what is beyond its bounds does not make sense. If we really had to, we could say that there is nothing there, but this leads – philosophers, at least – straight to the question of what this nothing is. Physics teaches us that it is by no means an envelope of that which is. And let it remain like this.

The deceptiveness of the language in which we make statements about nothingness is evident on the grounds of theology. God created the world out of nothing. This has to be absolute nothingness, for otherwise there would be something out of which the world was created, which would restrict God. We can hardly understand, however, what is meant by nothingness here if the term is used in the context of God's existence. How shall we express this? If we say that 'before' (and already this is difficult to verbalize, for it means a temporal relation, whereas theology, at least Christian theology, claims that God created time simultaneously with the world), which implies that before the creation there already existed God and nothingness, we contravene the grammar of the term 'exist'. If we say that there was nothing except God, we return to the dilemma noted by the antagonists of Paul of Tarsus. While we can always refer to the dogma of *creatio ex nihilo*, the coherence of a language so generated will be only illusory.

It is not entirely clear whether *Nichts* (*Néant*), as construed by Heidegger or Sartre, is absolute as defined by Parmenides or Leibniz. When Heidegger says that *Nichts* is pure non-being (he loved hyphenated words), he seems to follow Parmenides' footsteps, but when he argues that Nothingness nihilates, we can seriously doubt whether pure non-being acts in any way or is given originally in dread, whatever it is. According to the naturalistic conception, which I consider true, it is not Nothingness that nihilates *Dasein* but the – very much empirical – biological processes of ageing and the mental realization of the fact. From this point of view, dread is nothing but a fear of death and, therefore, an experience (or instinct, if you will) statistically common in the organic world. Perhaps, the specificity of being-for-itself, that is, human being (*Dasein*), makes it legitimate to speak about metaphysical dread, but – contrary to Heidegger again – this is not an ontological issue but a philosophical-anthropological one (Sartre was more of an anthropologist than an ontologist on this matter). Bound up with the philosophy of death, Heidegger's and Sartre's metaphysics of nothingness finds its justification in the singularity of the fact of the termination of life is certainly philosophically inspiring. Sartre was right to say (if I remember well, he was quoting someone else) that death is the only thing no one will do for us. There is no reason, therefore, to negate the gravity of it leading, as it does, to philosophical speculations. On the other hand, Epicurus' famous saying that 'When we exist, death is not; and when death exists, we are not' encapsulates, without unnecessary grandiloquence, the fact that death is an irreversible termination of life. But while it is true that death terminates the existence of a particular human being, this does not mean its transfer from Being to Nothingness. Please note how capital letters help to explicate certain theses. Understanding the imagery of Being and Nothingness becomes much easier when these terms are capitalized.

I do not intend to appeal for the category of nothingness to be dropped. But it is hard to accept the nothingness as construed by Parmenides and Leibniz holds any special significance for ontology. It is possible to speak about nothingness, though, when we refer it to a specific state of a fragment of being. Let us call such nothingness relative. A certain mathematical analogy can come in handy here. Some mathematical notions, namely the number zero (0) and an empty set ($\varnothing$), are regarded as illustrations of nothingness. This is a misconception. The number zero was introduced to make certain arithmetical operations universally executable. For example, for the difference between two numbers to be always definite, it was necessary to assume that $n - n = 0$. The next step was to assume that $n \cdot 0 = 0$. Still, division by zero is not possible. The same applies to an empty set. Let us assume that X and Y are sets of any value. Their intersection $X \cap Y$ is defined as the set of all those objects that belong to the set X and the set Y, so, for example, the intersection of the set of students and the set of athletes is the set of those students who are also athletes. But what do we do when the sets X and Y do not overlap, e.g. the set of cats and the set of dogs? We want the intersection of any two sets X and Y to be definite. To achieve this, we assume that $X \cap Y = \varnothing$, if X and Y share no elements. 0 and $\varnothing$ evoke certain associations with nothingness. For example, when I want to say that I have repaid my debt to Mr X, I may put it so that our balance is zero. The size of a debt may be expressed as a negative number and a positive balance as a higher-than-zero number, while zero expresses the lack of either a positive or negative balance. An even stronger sense of nothingness is evoked by an empty set, that is, a set containing no elements. On the other hand, 0 and $\varnothing$ are certain objects: 0 is a number and $\varnothing$ is a set. According to the aforementioned traditional view, if something (a being) exists, it is such and such. In this 0 and $\varnothing$ are beings. If they illustrate nothingness, it is only relatively, in relation to other numbers of non-empty sets. Their relativity consists in that they are not beyond being but are like holes or gaps in being. They are inscribed in being (the mathematical reality in the examples above) as its special places. They also fulfil various theoretical functions; for example, an empty set can be used to define natural numbers (the so-called von Neumann numbers): assuming ∅ = 0, {∅} = 1 ({∅} means 'a set whose only element is the set ∅), {∅, {∅}} = 2, {∅, {∅, {∅}, {Æ, {Æ}}} = 3 etc. Some mathematicians jokingly say that von Neumann's numbers are the only known example of a *creatio ex nihilo*, that is, creation out of nothing, out of an empty set. This, however, is but a metaphor with no mathematical meaning whatsoever. In reality, an empty set has such properties that by iterating constructions based on it we can define a sequence of natural numbers.

The notion of relative nothingness is encountered all the time. In 1993, I visited New York and viewed the World Trade Center towers. In 2009, I went to the same place and said, 'This is where the WTC towers used to stand and now there is nothing, except perhaps for a heap of rubble'. But it is also the other way round. When I came to New York in 2013, I found that a new building had been constructed at the former WTC site and the place was no longer the unbuilt vacuum it had been since September 2001. In *4'33"*, a piece by the 20th-century composer John Cage, the soloist and the orchestra do nothing for 4 minutes and 33 seconds. In this case, we are tempted to speak of a 'performance' rather than a performance. The score is a standard five-line, but the only signs in it are horizontal strokes. Whereas an ordinary music piece consists of sounds, *4'33"* consists of silence, that is, the lack of any sound. Aesthetically, the piece is a radical exemplification of an attitude known as *amor vacui* (love of void, emptiness). A similar attitude may have been expressed by Tadeusz Kantor when he exhibited – as artistic works – umbrellas with a sewn-on patch of cloth. The opposite aesthetics, which does not tolerate empty spaces – the arabesque is one example – is referred to as *horror vacui*, the same name that is given to the attitude that rejects the notion of an absolute vacuum in nature. One can imagine various degrees of plenum (a term denoting the opposite of empty space) and vacuum construed relatively. Cage, for example, could have placed one or more chords in *4'33"*, say, at 2'16" or any other moment. Kantor could have sewn on more patches of cloth. Reverse procedures are also possible. But without introducing suitable conventions, the question of when vacuum turns into plenum (or vice versa) cannot be answered. The cases considered so far are examples of reversible processes. Cage's vacuum can turn into a musical plenum, a *horror vacui* visual work can be transformed into an empty space. But there are cases when no reversibility is possible, not only because suitable reverse constructions have not been undertaken but by the very nature of nihilation processes consisting, for example, in an inevitable rise in entropy, making reversion impossible.

Relative vacua are bound up with certain ontological questions. Let us take, for example, the hole as a kind of void (sse Casati, Varzi 1994 for analysis of holes as specific objects). We may ask whether it has any structure, whether its boundaries belong to it or are part of the surroundings. Similar issues can be pondered with regard to other kinds of vacuum. Let us assume that every year Cage would have written a piece of finite length, but always a minute longer than the present one. Would then the structure of, say, *6'33"* be the same as that of *4'33"*? Relative vacua are empirical objects and each of them illustrates a general notion of relative nothingness, although sometimes the problem arises of whether we still have to do with a vacuum or already with a plenum (or the other way round). If we take two relative vacua, e.g. Kantor's umbrella with three sewn-on patches and Kantor's umbrella with a single patch, we are able to say that the latter better approximates Kantor's vacuum in the shape of an umbrella without any patch at all. Similarly, the vacuum of Cage's *4'33"* filled with sound in 33 percent better approximates normal musical plenum than the same vacuum filled in just 10 percent. It is actually possible that Cage would have protested calling *6'33"* (or a silence of any duration other than 4'33") 'Cage's vacuum'. He might insist that the duration of precisely 4 minutes and 33 seconds is what determines the piece's special aesthetic value. That relative *vacua* can remain in approximative or illustrative relations is not very surprising, for they are fully legitimate empirical objects, possessing quite a definite structure and such boundaries. Even an empty set is a mathematical structure, although a peculiar one, reduced to an empty relation. In any case, relative nothingnesses differ fundamentally from absolute nothingness. The latter, let me stress once again, has neither a structure nor boundaries, because it does not exist.

In the last two paragraphs, the term 'nothingness' was used only four times, in each case technically. I deliberately used various equivalents, e.g. vacuum, 'nothing', 'silence', 'hole', 'void' or 'emptiness'. In fact, specific cases of relative nothingness can be analysed without using the notion of nothingness at all. We will actually feel a certain sense of impropriety in saying that the World Trade Center was a nothingness after the September 11 attack and ceased to be one when a new building had been constructed there. Similarly, Cage's *4'33"* is certainly not nothingness either. It is not only that the remnants of the WTC towers were something, namely, a heap of rubble, and that Cage's piece is just silence during a normal concert gala (a soloist, an orchestra, a conductor, an audience), but that when we use the term 'nothingness', we automatically, as it were, attach to it all kinds of metaphorical meanings, be them theological or philosophical, as in the *creatio ex nihilo* narrative or the idea that Nothingness nihilates. It is possible that Heidegger preferred nihilation to annihilation, for the latter encompasses a broad range of situations, from the pacification (annihilation) of a worker strike to the murdering (annihilation) of several million people, whereas nihilation has a single predefined (though not necessarily clear) meaning. In most of its manifestations, relative nothingness is so prosaic that the very term 'nothingness' does not seem to apply. We may perhaps say that the *Titanic* with its passengers and crew ultimately sank into nothingness, but it is questionable whether saying that someone has plunged into nothingness because they drowned in the depths properly conveys what happened to that person. Since this is no place for analysing such cases in greater detail, I will continue avoiding (though not without exceptions) the term 'nothingness' and use vacuum (and its various synonyms) instead, always bearing in mind that the meaning of the latter is exactly relative, not absolute.

The Keret House in Warsaw can be analysed using the category of vacuum. A narrow crevice (void, gap, empty space) between the buildings of 22 Chłodna Street and 74 Żelazna Street inspired the designers of this architectural structure. Thus a certain vacuum has been transformed into a precisely defined plenum. Although the Keret House is touted as the world's narrowest living space, its position in the global ranking of narrow homes is utterly insignificant from the viewpoint of its role in demonstrating that even a very small empty space can be tightly filled. One could probably find a great many similar narrow clearings elsewhere, doubtlessly even such that are bound, on one side, by a relatively old tenement and, on the other, by a relatively modern residential block housing numerous flats or, as they are called today, apartments. The technical parameters of the Keret House can be very interesting from the viewpoint of architectural craftsmanship and the mode of turning vacuum into plenum; for example, we could inquire, in the vein of Roberto Casati and Achille Varzi (see above), whether the boundaries of a former vacuum automatically become the boundaries of the new plenum. But this does not determine the question I consider in this essay. I view the Keret House as something very special, not due to its architectural values, but to the possible (for not everyone has to think the same) symbolic meaning inscribed in its historical context.

Let us assume that the structure installed in the gap between 22 Chłodna and 74 Żelazna Streets would have a different name, e.g. the Narrow House. But if the casual passer-by remembers that the Narrow House is located in this and not another place (the Ghetto, the Bridge of Sighs), they may wonder whether this particular plenum can be interpreted as a token of something that is unconnected with its present-day function. The results of such reflection could be diverse. Someone, let us call him Mr X, might think that the clearing between the buildings should have been left empty as a symbol of the annihilation of the Jews, and perhaps add that, in a way, it is this void. Moreover, if that person knew the writings of Heidegger and Sartre, he or she might add that the view of the streets separating the two parts of the Ghetto, as seen from the footbridge connecting them, would illustrate quite aptly the original nothingness given in dread, for these streets, in fact full of hustle and bustle, illustrated the nihilating border between life and death. On the other hand, someone else, a Mr Y, might conclude that the filling of a specific vacuum by the Narrow House, that is, a specially designed plenum, is very much legitimate and epitomizes the re-being, re-existence or revitalisation of what was once destroyed with the intention of total annihilation. The geometric shape of the original vacuum and the architectural characteristics of the created plenum are of secondary importance. The symbolism has to do with the difference between the world of life and the world of death, viewed, as it is, from two different directions. The attitudes of Mr X and Mr Y towards the Keret House reflect a dualism in its interpretation, depending on whether the perspective is past- or future-oriented.

But one might add here one more perspective. The Warsaw Ghetto was doubtlessly something close to absolute nothingness given in original dread, to use Heidegger's terminology, albeit indexing wholly empirical phenomena. And this is symbolized by the gap between the buildings at the corner of Chłodna and Żelazna. But life in the Ghetto was not limited to being subject to nihilation, whatever this might mean. One could also (leaving aside the daily affairs and clandestine activity) graduate from high school there, complete three years of medicine studies, research the phenomenon of hunger (on a particularly large sample), listen to concerts, compose music, write poems and novels, play in theatres and go to them, run orphanages (like Janusz Korczak) or assemble historical archives (like Emanuel Ringelblum), that is, do all that can be done in ordinary circumstances. Perhaps, not in conditions as comfortable as those provided by the Keret House today, but on no small a scale nonetheless. It does not make sense, of course, to put these two aspects of living in the Ghetto on a par. It was the topsy-turvy world that dominated, with its life-annihilating standards, rather than the artistic attempts to fill the ever-greater vacuum. Nor did the world described by Ka-Tzetnik consist solely in inertly plunging into a vacuum governed by special 'breath-regulating' laws, although it was much closer to liminal nothingness, as it were, than the reality of the Jewish housing districts. Let the Keret House, therefore, symbolize not only the chasm between the world of life and the world of death, but also the attempts, both heroic and not, to cross the boundary.

Bibliography

Aberdein, A. 2008, "Logic for Dogs" in Hales 2008, 167–181.

Aertsen, J. A. 1996, *Medieval Philosophy & the Transcendentals. The Case of Thomas Aquinas*, Brill, Leiden,

Aertsen, J. A. 2012, *Medieval Philosophy as Transcendental Thought: From Philip the Chancellor (ca. 1225) to Francisco Surarez*, Brill, Leinden.

Agazzi, E. 2004, *Right, Wrong and Science. The Ethical Dimensions of the Techno-Scientific Enterprise*, Rodopi, Amsterdam.

Agazzi, E. (ed.) 1981, *Modern Logic*, Reidel, Dordrecht.

Ajdukiewicz K. 1955, "Klasyfikacja rozumowań" (Classification of Reasoning), *Studia Logica* II, 278–300.

Andrews, P. B. 1986, *An Introduction to Mathematical Logic and Type Theory: To Truth from Proof*, Academic Press, Orlando.

Antonelli, G. A. 2000, "Virtuous Circles: From Fixed, Points to Revision Rules", in Chapuis, Gupta 2000, 1–27.

Araszkiewicz, M., Banaś, P., Gizbert-Studnicki, T., Płeszka, K. (eds.) 2014, *Problems of Normativity, Rules and Rule-Following*, Springer, Berlin.

Armstrong, D. 2004, *Truth and Truthmakers*, Cambridge University Press, Cambridge.

Arrigoni, T. 2007, *What is Meant by V? Reflections on the Universe of All Sets*, Mentis, Paderborn.

Asser G. 1972, *Einführung in die mathematische Logik, Teil II, Prädikatenkalkül der ersten Stufe*, Teubner, Leipzig.

Auletta G. 2011, *Cognitive Biology. Dealing with Information from Bacteria to Mind*, Oxford University Press: Oxford.

Badesa, C. 2004, *The Birth of Model Theory. Löwenheim's Theorem in the Frame of the Theory of Relatives*, Princeton University Press, Princeton.

Barbieri M. 2003, *The Organic Codes. An Introduction to Semantic Biology*, Cambridge University Press: Cambridge.

Barbieri M. (ed.) 2008, *Introduction to Biosemiotics. The New Biological Synthesis*, Springer, Berlin.

Barke, A. 2002, *The Closure of Knowledge in Context*, Mentis, Paderborn.

Barnes, J. 1979, *The Presocratic Philosophers*, Routledge, London.

Bartlett, S. J. (ed.) 1992, *Reflexivity. A Source-Book in Self-Reference*, North-Holland, Amsterdam.

Bartlett, S. J., Suber, P. (eds.) 1987, *Self-Reference. Reflections on Reflexivity*, Nijhoff, The Hague.

Barwise, J., Etchemendy, J. 1987, *The Liar: An Essays in Truth and Circularity*, Clarendon Press, Oxford.

Barwise, J., Moss, L. 1996, *Vicious Circle*, CSLI Publications, Stanford.

Bates A., Maxwell A. 2005, *DNA Topology*, Clarendon Press, Oxford.

Beebee, H. Dodd, J. (eds.) 2005, *Truthmakers. The Contemporary Debate*, Clarendon Press, Oxford.

Beeson, M. J 2004, *Foundations of Constructive Mathematics*, Springer, Berlin.

Bell, J. L. 1999, *The Art of Intelligible. An Elementary Survey of Mathematics and its Conceptual Development*, Kluwer, Dodrecht.

Bernstein, M. H. 1992, *Fatalism*, University of Nebraska Press, Lincolm.

Betti, A. 2002, "The Incomplete Story of Łukasiewicz and Bivalence", in *The Logica Yearbook 2001*, Filosofia, Praha, 21–36.

Béziau, J.-Y. 2004, "Bivalence, Excluded Middle and Non-Contradiction", in *The Logica Yearbook* 2003, Filosofia, Praha, 75–83.

Béziau, J.-Y. (ed.) 2007, *Logica Universalis. Toward a General Theory of Logic*, 2nd ed., Birkhäuser, Basel.

Béziau, J.-Y., Carnielli, W., Gabbay, D. (eds.) 2007, *Handbook of Paraconsistency*, College

Publications, London.

Black, M. 1948, "The Semantic Definition of Truth", *Analysis* 9, 49–63; repr. in Black 1949, 89–107.

Black, M. 1949, *Language and Philosophy*, Cornell University Press, Ithaca.

Bollander, T., Hendricks, V. T., Pedersen 2006, "Introduction", in Bollander, T., Hendricks, V. T., Pedersen, S. 2006, 1–25.

Beth, E. W. 1962, *Formal Methods. An Introduction to Symbolic Logic and to the Study of Effective Operations*, D. Reidel, Dordrecht.

Blackburn, P., Van Benthem, J., Wolter, F. (eds.) (2007), *Handbook of Modal Logic*, Elsevier, Amsterdam.

Blanché, R. 1953, "Sur le opposition des concepts", *Theoria* 19, 89–130.

Bolander, T., Hendricks, V. T., Pedersen, S. (eds.) 2006, *Self-Reference*, CSLI Publications, Stanford.

Bonevac, D. 1990, *The Art and Science of Logic*, Mayfield, Mountain View.

Borkowski, L. 1991, "Some Remarks on the Principle of Bivalence and Many-valued Logics", in Borkowski, Stępień 1991, 25–30.

Borkowski, L., Stepień, A. (eds.) 1991, *Studies in Logic and Theory of Knowledge 2*, Towarzystwo Naukowe Katolickiego Uniwersytetu Lubelskiego, Lublin 1991.

Botha R. 2003, *Unravelling the Evolution of Language*, Elsevier: Amsterdam.

Bradbury J., Vehrencamp, S. 1998, *Principles of Animal Communication*, Sinauer, Sunderland, Mass.

Bridges, D., Richman, F. 1987, *Varieties of Constructive Mathematics*, Cambridge University Press, Cambridge.

Bromand, J. 2001, *Philosophie der semantischen Paradoxien*, Mentis, Paderborborn.

Brooks D., Wiley E. 1986, *Evolution as Entropy. Toward a Unified Theory of Biology*, University of Chicago Press: Chicago.

Brożek, B. 2012. *Rule-Following. from Imitation to the Normative Mind.* Copernicus Center Press, Kraków.

Burris, S. 1998, *Logic for Mathematics and Computer Science*, Prentice-Hall, Upper Sadle River, NJ.

Cahn, S. M. 1967, *Fate, Logic and Time*, Yale University Press, New Haven.

Caleiro, C., Carnielli, W., Coniglio, M. M., Marcos, J. 2007, "Two's Company: 'The Humbug of Many Logical Values'", in Béziau, 2007, 175–194.

Cann, R., 1993, *Formal Semantics. An Introduction*, Cambridge University Press, Cambridge.

Chappel, V. (ed.) 1964, *Ordinary Language*, Prentice-Hall, Englewood-Clifts:

Chapuis, A., Gupta. A. (eds.) 2000, *Circularity, Definition and Truth*, Indian Council of Philosophical Research, New Delhi.

Carnap R. 1936, "[Diskussion]", in *Actes du Huitième Conqrès International de Philosophie à Prague 2–7 Septembre 1934*, Orbis, Prague, 244.

Carnap, R. 1947, *Meaning and Necessity. A Study in Semantics and Modal Logic*, University of Chicago Press, Chicago.

Casati, R., Varzi, A. 1994, *Holes and Other Superficialities*, The MIT Press, Cambridge, Mass.

Childers, T., Majer, O. (eds.) 1995, *Logica'94. Proceedings of the 8th International Symposium*, Filosofia, Praha 1995.

Church A. 1956, *Introduction to Mathematical Logic*, Princeton University Press, Princeton.

Code, L. 1987, *Epistemic Responsibility*, University Press of New England, Hannover.

Cooper W. 2001, *The Evolution of Reason. Logic as a Branch of Biology*, Cambridge University Press, Cambridge.

Dégremont, C., Keiff, L., Rückert, H. (eds.) 2008, *Dialogues, Logics abd Other Strange Things. Essays in Honour of Shahid Rahman*, College Publications, London.

Dehaene S. 1997, *The Number Sense. How the Mind Creates Mathematics*, Clarendon Press, Oxford.

Dragalin, A. G. 1987, *Mathematical Intuitionism. Introduction to Proof Theory*, American Mathematical Society, Providence, Rhode Island.

Duda R. 1986, *Wprowadzenie do topologii*, Vol. 1: *Topologia ogólna* (Introduction to Topology, Vol. 1: General Topology), Państwowe Wydawnictwo Naukowe, Warszawa.

Dummett, M. 2000, *Elements of Intuitionism*, 2nd ed., Clarendon Press, Oxford.

Dutillh Novaes, C. 2008, "Tarski's Hidden Theory of Meaning: Sentences Say Exactly", in Rahman, Tulenheimo, Genot 2008, 41–63.

Ebbesen S. 1980, *Logica docens/logica utens*, in Ritter and Gründer 1980, 353–355.

Ebbs, G. 1997. *Rule Following and Realism*, Harvard University Press, Cambridge, Mass.

Edwards, P. (ed.) 1967, *The Encyclopedia of Philosophy*, ed. by P. Edwards, Vol. 5, Macmillan, New York.

Edwards, P., Pap, A. (eds.) 1957, *A Modern Introdtuction to Philosophy. Readings from Classical and Contemporary Sources*, The Free Press of Glencoe, New York.

Elders, L. J. 1992, *The Metaphysics of Being of St. Thomas Aquinas in a Historical Perspective*, Brill, Leiden.

Enderton, H. 1972, *A Mathematical Introduction to Logic,* Academic Press, New York.

Ewing A. C. 1937, "Meaninglesness", *Mind* XLVI, 40–49 ; repr. in Edwards, Pap 1957,770–779.

Favareau D. (ed.) 2009, *Essential Readings in Biosemiotics. Anthology and Commentary*, Springer: Berlin.

Flint, Th., Rea, M. (ed.) 2009, *The Oxford Handbook of Philosophical Theoology*, Clarendon Press, Oxford.

Fraenkel, A., Bar-Hillel, Y., Levy, A. 1973, *Foundations of Set Theory*, North-Holland, Amsterdam.

Frege, G. 1879, *Begriffschrift, eine der arithmetischen nachgebildete Formelsprache des reinen Denkens*, Nerbert, Halle 1879; Eng. tr. In Van Heijenoort 1967a, 5–82.

Frege, G. 1879–1891, „Logik", in Frege 1969, 1–8; Eng. tr. in Frege 1979, 1–9.

Frege, G. 1897, *Logik*, in Frege 1969, 137–163; Eng. tr. in Frege 1979, 126–151.

Frege, G. 1918–1919, *Der* Gedanke, *Beiträge zur Philosophie der deutschen Idealismus* I, 58–77; Eng. tr. in Frege 1984, 351–372.

Frege, G. 1969, *Nachgelassene Schriften*, Meiner, Hamburg.

Frege, G. 1979, *Posthumus Writings*, Blackwell, Oxford.

Frege, G. 1984, *Collected Papers on Mathematics, Logic and Philosophy,* Blackwell, Oxford.

Gabbay, D., Guenther, F. (eds) 1984, *Handbook of Philosophical Logic*, Vol. III, *Extensions of Classical Logic*, D. Reidel, Dordrecht.

Gabbay D., Woods, J. (eds.) (2006), *Handbook of the History of Logic*, Vol. 7, *Logic and the Modalities in the Twentieth Century*, Elsevier, Amsterdam.

Gaifman, H. 1992, "Pointers to Truth", *The Journal of Philosophy* 89, 223–262.

Gaifman, H. 2000, "Pointers to Propositions", in Chapuis, Gupta 2000, 79–121.

Garson, J. W. 2013, *What Logics Mean. From Proof Theory to Model-Theoretic Semantics*, Cambridge University Press, Cambridge.

Gaskin, D. 1995, *The Sea Battle Argument and the Master Argument. Aristotle and Diodorus*

Cronus, de Gruyter, Berlin.

Gottwald, S. 2001, *A Treatise on Many-Valued Logic*, Research Studies Press, Baldock.

Gödel, K. 1934. *On Undecidable Propositions of Formal Mathematical Systems* (mimeographed notes), Princeton: Princeton University Press; repr. (with additions) in Gödel 1986, 346–371.

Gödel, K. 1986. *Collected Works*, Vol. I, Clarendon Press, Oxford.

Greco, J., Turri, J. 2011, "Virtue Epistemology", in *The Stanford Encyclopedia of Philosophy*, ed. by E. Zalta, http://plato.stanford.edu/archives/win2011/entries/epistemology-virtue/.

Greniewski, H. 1970, *Sprawy wszystkie i jeszcze inne* (All Matters and Still More), Książka i Wiedza, Warszawa.

Greniewski, H. 1970a, "2^{n+1} wartości logicznych" (2^{n+1} of Logical Values), in Greniewski 1970, 127–201.

Grzegorczyk, A. 1974, *An Outline of Mathematical Logic. Fundamental Results and Notions Explained with All Details*, Reidel, Dordrecht.

Grzegorczyk A. 1997, *Logic – A Human Affair*, Scholar, Warszawa.

Gupta, A. 2000, "On Circular Concepts", in Chapuis, Gupta 2000, 123–153.

Haack, S. 2002, "The Ethics of Belief Reconsidered", in Steup 2002, 21–33.

Hales, S. D. (ed.) 2008, *What Philosophy Can Tell You about Your Dog*, Open Court: LaSalle.

Haack, S. 2002, "The Ethics of Belief Reconsidered", in Steup 2002, 21–33.

Halpern, J. Y. (ed.) 1986, *Reasoning about Knowledge. Proceeding of the 1986 Conference*, Morgan Kaufmann, Los Altos.

Hansson, S. O. 2001, *The Structure of Values and Norms*, University of Cambridge Press, Cambridge.

Hauser M. 1998, *The Evolution of Communication*, The MIT Press, Cambridge, Mass.

Heidegger, M. 2014, *What Is Metaphysics?*, trans. M. Groth, http://wagner.wpengine.netdna-cdn.com/psychology/files/2013/01/Heidegger-What-Is-Metaphysics-Translation-GROTH.pdf [retrieved 23 January 2014].

Heine B., Kuteva, T. 2007, *The Genesis of Grammar. A Reconstruction*, Clarendon Press, Oxford.

Heller, M. 1999, "Przeciw fundacjonizmowi" (Against Foundationism), in Heller, Mączka, Urbaniec 1999, 81–101.

Heller, M., Mączka, J., Urbaniec, J. (eds.) 1999, *Sensy i nonsensy w nauce i filozofii* (Senses and Nonsenses in Science and Philosophy), OBI-Biblos. Kraków-Tarnów.

Hendricks, V. F., Pedersen, S. A., K. F. Jørgensen, K. F. (eds.) 2000, *Proof Theory. History and Philosophical Significance*, Kluwer, Dordrecht.

Hendricks, V. F., Neuhaus, F., Pedersen, S. A., Scheffler, U., Wansing, H. (eds.) 2004, *First-Order Logic Revisited*, λογος, Berlin.

Herbrand, J. 1971, *Logical Writings*, Reidel, Dordrecht.

Heyting, A. (ed.) 1957, *Constructivity in Mtghematics*, North-Holland, Amsterdam.

Hilbert, D., Ackermann, W. 1928, *Grundzüge der theoretischen Logik*, Julius Springer, Berlin; Engl. tr. Chelsea, New York 1950.

Hilbert, D., Bernays, P. 1934–1939. *Grundlagen der Mathematik.* Berlin: Julius Springer.

Hinman P. G. 2005, *Fundamentals of Mathemathical Logic*, A. K. Peters, Wellesley.

Hintikka, J. 1988, "On the Development of the Model-Theoretic Viewpoint in Logical Theory", *Synthese* 77, 1–36; repr. in Hintikka 1997, 104–139.

Hintikka, J. 1989, “Is Truth Ineffable?”, in Scardona 1989, 89–120; repr. in Hintikka 1997, 20–45.

Hintikka, J. 1997, *Lingua Universalis vs. Calculus Ratiocinator: An Ultimate Presupposition of Twntieth-Century Philosophy*, Kluwer, Dorrecht.

Hintikka, J. (in collaboration with U. Remes and S. Knuuttila) 1977, *Aristotle on Modality and Determinism*, North-Holland, Amsterdam.

Hochberg, H. 1984, *Logic, Ontology and Language. Essays on Truth and Language*, Philosophia Verlag, München.

Hochberg, H. 1984a, “Facts and Truth”, in Hochberg 1984, 279–295. *Logic, Ontology and Language. Essays on Truth and Language*, Philosophia Verlag, München,

Hoffmeyer J. 2008, *Biosemiotics. An Examination into the Signs of Life and the Life of Signs*, University of Scranton Press, Scranton.

Horsten, L, Pettigrew, R. (eds.) (2011), *The Continuum Handbook of Philosophical Logic*, Continuum, London.

Hume, D. 1951, *A Treatise on Human Nature*, Clarendon Press, Oxford (first published in 1739).

Hurford J. 2007, *The Origins of Meaning. Language in the Light of Evolution*, Clarendon Press, Oxford.

Hurford J. 2012, *The Origin of Grammar. Language in the Light of Evolution*, Clarendon Press, Oxford.

Ilachinski A. 2011, *Cellular Automata. A Discrete Universe*, World Scientific, Singapore.

Ingarden, R. 1935, “L'essai logisttque d'une refonte de la philosophie”, *Revue Philosophique* 120, 137–159.

Ingarden, R. 1936, “Der Iogtstische Versuch einer Neugestaltung der Philosophie. Eine kritische Bemerkung“, *Actes du Huitième Conqrès International de Philosophie à Prague 2–7 Septembre 1934*, Orbis, Prague, 203–208.

Jadacki, J., Paśniczek, J. (eds.) 2006, *The Lvov-Warsaw School – The New Generation*.,Amsterdam: Rodopi.

Johannson, L. G., Österberg, J., Śliwiński, R. (eds.) 2009, in: *Logic, Ethics, and All That Jazz. Essays in Honour of Jordan Howard Sobel*, Uppsala Universitet, Uppsala.

Johansson S. 2005, *Origins of Language. Constraints on Hypotheses*, John Benjamins, Amsterdam.

Jordan, Z. 1963, “Logical Determinism”, *Notre Dame Journal of Formal Logic* 4, 1–38.

Karpenko, A. 1990, *Fatalism i słuczajnost' buduscevo. Łogiceskij analiz* (Fatalism and Contingency of the Future. A Logical Analysis; in Russian), Nauka, Moskva.

Kauffman S. 1993, *The Origin of Order. Self-Organization and Selection in Evolution*, Clarendon Press, Oxford.

Kay L. 2000, *Who Wrote the Book of Life? A History of the Genetic Code*, Stanford University Press, Stanford.

Kerferd, G. B. 1967, "Nothingness", Edwards 1967, 324.

Klug W., Cummings M., Spencer Ch. 2006, *The Concepts of Genetics*, Prenticed-Hall, Upper Saddle River, NJ.

Kneebone, G. 1963, *Mathematical Logic and the Foundations of Mathematics*, Van Nostrand, New York.

Knittermeyer, H. 1920, *Der Terminus "transzendental" in seiner historischen Entwicklung bis zu Kant*, Hamek, Marburg.

Kokoszyńska M. 1936, "Über den Absoluten Wahrheitsbegriff und einige andere semantische Begriffe", *Erkenntnis* 6, 143–165; repr. in Pearce, Woleński 1988, 276–292.

Kokoszyńska M. 1948, "What Means a „Relativity" of Truth", *Studia Philosophica* III, 167–175.

Kokoszyńska M. 1951, "A Refutation of the Relativism of Truth", *Studia Philosophica* IV, 93–149.

Kolak, D., Symons, J. (eds.) 2004, *Quantifiers, Questions and Quantum Physics*, Berlin.

Koons, R. 2000, "Circularity and Hierarchy", in Chapuis, Gupta 2000, 117–197.

Kotabiński, T. 1913, "Zagadnienie istnienia przyszłości" (The problem of the Existence of the Future), *Przegląd Filozoficzny* 16, 74–92; Eng. tr., *The Polish Review* 13(3) (1968), 7–22.

Kripke, S. 1971, "Identity and Necessity", Munitz 1971, 135–164; repr. in Kripke 2011, 1–26.

Kripke, S. 1975, "Outline of an Theory of Truth", *The Journal of Philosophy* 72, 690–712.

Kripke, S. 1982, *Wittgenstein on Rules and Private Language*. Harvard University Press, Cambridge, Mass.

Kripke, S. 2011, *Philosophical Troubles. Collected Papers*, Vol. 1, Clarendon Press, Oxford.

Kusch, M. 1989, *Language as Calculus vs. Language as Universal Medium. A Study on Husserl, Heidegger and Gadamer*. Kluwer, Dordrecht.

Küppers B.-O. 1990, *Information and the Origin of Life*, The MIT Press, Cambridge, Mass.

Lamm E., Unger R. 2011, *Biological Computation*, CRC Press, London.

Langer J. 1980, *The Origins of Logic. Six to Twelve Months*, Academic Press, New York.

Larson R., Déprez Y., Yamakido H. (eds.) 2010), *The Evolution of Human Language. Biolinguistic Perspective*, Cambridge University Press, Cambridge.

Leśniewski, S. 1912, "Czy prawda jest tylko wieczna, czy też wieczna i odwieczna? (Is all Truth only True Eternally or It Is True without a Beginning), *Nowe Tory*, 493–512; Eng. tr. in Leśniewski 1992, 86–114.

Leśniewski, S. 1931, "O podstawach matematyki" (On the Foundations of Mathematics) §§10-11, *Przegląd Filozoficzny* 34, 142–170; Eng. tr. in Leśniewski 1992, 351–382.

Leśniewski, S. 1992, *Collected Works*, Kluwer, Dordrecht.

Lewis, H. (ed.) 1956, *Contemporary British Philosophy*, III. George Allen and Unwin, London,

Lieberman Ph. 2005, *Toward an Evolutionary Biology of Language*, Harvard University Press, Cambridge, Mass.

Lowe, E. J., Rami, A. (eds.) 2009, *Truth and Truth-Making*, Acumen, Stocksfields.

Löb, M. 1955, "Solution of the Problem of Leon Henkin", *Journal of Symbolic Logic* 20, 115–118.

Lucas, J. R 1989, *The Future. An Essay on God, Temporality and Truth*, Blackwell, Oxford.

Luisi P. 2006, *The Emertgence of Life. From Chemical Origins to Synthetic Biology*, Cambridge University Press, Cambridge.

Łukasiewicz, J. 1922, "On Determinism" in Łukasiewicz 1970, 110–128.

Łukasiewicz, J. 1930, "Philosophische Bemerkungen zu mehrwertigen Systemem der Aussagenkalkül", *Comptes Rendus des Sèances de la Societé et de Lettres de Varsovie, Classe III* 23, 51–77; Eng. tr. in Łukasiewicz 1970, 153–178.

Łukasiewicz, J. 1970, *Selected Works*, North-Holland, Amsterdam.

Malinowski, G. 1993, *Many-Valued Logic*, Clarendon Press, Oxford.

Martin, R. L. (ed.) 1978, *The Paradox of the Liar*, Ridgeview, Reseda.

Martin, R. J. (ed.) 1984, *Recent Essays on Truth and the Liar Paradox*, Clarendon Press, Oxford.

Massey G. J., 1970, *Understanding Symbolic Logic*, Harper and Row, New York.

McCarty, D. 1994, "On Theorems of Gödel and Kreisel: Completeness and Markov's Principle", *Notre Dame Journal of Formal Logic* 35, 99–106.

McNamara P. 2006, "Deontic Logic, in Gabbay, Woods 2006, 197–288.

Mendelson, E. 1964, *Introduction to Mathematical Logic*, Van Nostrand, New York.

Miller, D. 2010, "Verdades, Antiguas Y Modernas". *Dicussiones Filosóficas 18*, 207–227.

Monmoyer, J.-M., (ed.) 2007, *Metaphysics and Truthmakers*, Ontos, Franfurt am Main.

Mostowski, A. 1957, "On Various Degrees of Constructivism, in Heyting 1957, 178–194.

Mostowski, A. 1965, *Thirty Years of Foundational Studies. Lectures on the Development of Mathematical Logic and the Foundations of Mathematics in 1930–1964*, Societas Philosophica Fennica, Helsinki.

Mulligan, K. 2007, "Two Dogmas of Truthmaking", in Monmoyer 2007, 51–65.

Mulligan, K., Simons, P. M., Smith, B. 1984, "Truth-Makers", *Philosophy and Phenomenological Research* 44, 287–321; repr. in Monmoyer 2007, 9–49.

Munitz. M. K. (ed). 1971, *Identity and Individuation*, New York University Press, New York Murawski, R. 1999, *Recursive Fuctions and Metamathematics. Problems of Completeness and Decidabilit, Gödel's Theorems*, Kluwer Academic Publishers, Dordrecht.

Nederpelt, R., Kamarredine, F. 2004, *Logical Reasoning: A First Course*, King's College, London.

Nef, F. 2007, "*Senex erit puer.* Truthmakers for Tensed Sentences", in Monmoyer 2007, 221–235.

Nelson Ph. 2008, *Biological Physics. Energy, Information, Life*, W. H. Freeman, New York.

Niiniluoto, I., Sintonen, M., Woleński, J. (eds.) 2004, *Handbook of Epistemology*, Kluwer, Dordrecht.

Novikov, P. S. 1977, *Konstructivnaia matemaceskaia logika iz tocki zrenija klasiceskoj logiki* (Mathematical Constructive Logic form the Point of View of Classical Logic), Nauka, Moskva.

Pap, A. 1957, "Introduction [to VIII. Part]", in Edwards, Pap 1957, 450–455.

Papineau D. 1993, *Philosophical Naturalism*, Blackwell, Oxford.

Parsons, J. 2005, "Truthmakers, the Past and the Future", in Beebee and Dodd 2005, 161-174.

Patterson, D. 2012, *Alfred Tarski Philosophy of Language and Logic*, Palgrave Macmillan, Hampshire.

Pearce, D., Woleński, J. (eds.) 1988, *Logische Rationalismus. Philosophische Schriften der Lemberg-Warschauer Schule*, Athenäum, Frankfurt am Main.

Peregrin, J. (ed.) 1999, *Truth and Its Nature (If Any)*, Kluwer, Dodrecht.

Perzanowski, J. (draft), "O wskazanych przez Hartshorne'a modalnych krokach w dowodzie ontologicznym Anzelma (1)" (On Modal Steps in Anselm's Ontological Proof (1)). http://www.tezeusz.pl/cms/tz/index.php?id=4839.

Perzanowski, *Art of Philosophy*, Ontos, Frankfurt am Main.

Perzanowski, J. 2012a, „Reasons of Monodeism", in Perzanowski, 411–414.

Placek, T. 2006: "A Puzzle about Semantic Determinism: Łukasiewicz's "On Determinism" – Thirty Years Laterin ", in Jadacki, Paśniczek 2006, 171–185.

Pogorzelski, W. A. 1994, *Notions and Theories of Elementary Formal Logic*, Warsaw University – Białystok Branch, Białystok.

Portner, P. 2009, *Modality*, Clarendon Press, Oxford.

Post, E. 1921, "Introduction to a General Theory of Elementary Propositions", *Americal Journal of Mathematics* 43, 163–185; repr. in Post 1994, 21–43.

Post, E. 1994, *Solvability, Provability, Definability: The Collected Works of Emil L. Post*. Birkhäuser, Basel.

Priest, G., Tanaka, K. 2009, "Paraconsistent Logic", *The Stanford Encyclopedia of Philosophy (Summer 2009 Edition)*, ed. by E. N. Zalta , URL = <http:// plato.stanford.edu/ archives/ sum2009/entries/logic-paraconsistent/>.

Priest, G.., Berto, F. 2010, "Dialetheism", in *The Stanford Encyclopedia of Philosophy (Summer 2010 Edition)*, Edward N. Zalta (ed.), URL=<http://plato. stanford.edu/archi- vcs/sum2010/cntries/dialetheism/>.

Prior, A. N. 1953, "Three-Valued Logic and Future Contingents", *The Philosophical Quarterly*, 317–326.

Rahman, S., Tulenheimo, T., Genot, E. (eds.) 2008, *Unity, Truth and the Liar. The Modern Relevance of Medieval Solutions to the Liar Paradox*, Springer, Dordrecht.

Rami, A. 2009, "Introduction: Truth and Truth-Making", in Loewe and Rami 2009, 1–36.

Read, S. 2008, "The Truth-Schema and the Liar", in Rahman, Tulenheimo, Genot 2008, 3–18.

Read, S. 2008a, "Further Thoughts on Tarski's T-scheme and the Liar", in Rahman, Tulenheimo, Genot 2008, 3–18.

Read, S. 2010, "The Liar and the New Truth-Schema", *Dicussiones Filosóficas 17*, 119–137.

Read, S. 2011, "Miller, Bradwardine and Truth", *Dicussiones Filosóficas 18*, 229–235.

Rescher, N. 1969, *Many-Valued Logic*, MGraw-Hill, New York.

Risse, W. 1980, *Logik*, in Ritter and Gründer 1980, 357–362.

Ritter, J., Gründer, K. (eds.) 1980, *Historischer Wörterbuch der Philosophie*, Vol. 5, Schwabe, Basel.

Rojszczak, A., Cachro, J., Kurczewski, G. (eds.) 2003, *Philosophical Dimensions of Logic and Science. Selected Contributed Papers from the 11th International Congress of Logic, Methodology, and Philosophy of Science*, Kraków, 1999, Kluwer, Dordrecht.

Rosser, J. B. 1953, *Logic for Mathematicians*, McGraw Hill, New York.

Russell, B. 1903, *The Principles of Mathematics*, Georg Allen and Unwin, London.

Russell, B. 1912, *The Problems of Philosophy*, Wiliam and Norgate, London.

Sandu, G. 2011, "Read on the Liar", *Dicussiones Filosóficas* 17, 277–290.

Sartre, J. P. 2014, *Being and Nothingness*, trans. H. E. Barnes, http://www.dhspriory.org/ kenny/ PhilTexts/ Sartre/BeingAndNothingness.pdf [retrieved 23 January 2014].

Scardona N. (ed.) 1989, *Les Formes Actuelles du Vrai: Entretiens de Palermo*, Enhiridion, Palermo.

Schlick, M. 1931, "Das Kausalität in den gegenwärtigen Physik", *Naturwisseschaften* 19,

145-162; Eng. tr. Schlick 1979, 176–209.

Schlick, M. 1979, *Collected Papers*, Vol. II, Reidel, Dordrecht.

Schmidt, J. 2003, *Philosophische Theologie*, Kohlhammer Verlag, Stuttgart.

Schulemann, G. 1929, *Die Lehre von den Trans- zendentalien in der scholistischen Philosophie*, Meiner, Leipzig.

Schulz, G. 1993, *Untersuchungen zur Wahrheitslehre des Thomas von Aquin & zur Kritik Kants Wahrheitsbegrif*, Brill, Leiden.

Sheard, M. 1994, "A Guide to Truth Predicates in the Modern Era", *Journal of Symbolic Logic* 59, 1032–1054.

Shoenfield, J. 1967, *Mathematical Logic*, Addison-Wesley, London.

Shramko, Y., Wansing, H. 2011, *Truth and Falsehood. An Inquiry into Generalized Logi-cal Values*, Springer, Dordrecht.

Sieg, W. 2000, "Toward Finitist Proof Theory", in Hendricks, Pedersen, Jørgensen 2000, 96–114.

Simons, P. 1986, "Łukasiewicz, Meinong, and Many-Valued Logic", in Szaniawski 1989, 249–291.

Smorynski, C. 1985, *Self-Reference and Modal Logic*, Springer, Berlin.

Smith, B. C. 1986, "Varietes of Self-Reference", in Halpern 1986, 19–43.

Smith, B., Simon, J. 2007, "Truthmaker Explanations", in Monmoyer 2007, 79-98.

Smullyan, R. M. 1992, *Gödel's Incompleteness Theorems*, Oxford University Press, Oxford.

Sobel, H. 2004, *Logic and Theism. Arguments for and against Beliefs in God*, Cambridge University Press, Cambridge.

Sommaruga-Rosolemos, G. 1991, *Fixed Point Constructions in Various Theories of Mathematical Logic*, Bibliopolis, Napoli.

Stam J. 1976, *Inquiries into the Origin of Language. The Fate of a Question*, Harper and Row: New York.

Steup, M. (ed.) 2002, *Konwledge, Truth and Duty. Essays on Epistemic Justification, Responsibility, and Virtue*, Clarendon Press, Oxford.

Surma, S. J. 1981, "The Growth of Logic out of the Foundational Research in the Foundations of Mathematics", in Agazzi 1981, 15–33.

Suszko, R., 1975, "Remarks on Łukasiewicz's three-valued logic", *Bulletin of the Section of Logic* 4, no. 3, 87-90.

Suszko, R. 1977, "The Fregean Axiom and the Polish Mathematical Logic in the 1920s", *Studia Logica* XXXVI, 376–380.

Swaart, H. C. de 1976, *Intuitionistic Logic in Intuitionistic Mathematics*, Faculteit der Wiskunde en Natuurwtenschappen, Nijmegen.

Szaniawski, K. (ed.) 1989, *The Lvov–Warsaw School and the Vienna Circle*, Nijhoff, The Hague.

Talerman, M., Gibson, R. (ed.) (2012), *The Oxford Handbook of Language Evolution*, Clarendon Press, Oxford.

Tarski, A. 1933, *Pojęcie prawdy w językach nauk dedukcyjnych* (The Concept of Truth in Languages of Deductive Sciences), Towarzystwo Naukowe Warszawskie, Warszawa; Germ. tr, Tarski 1935, Eng. tr. in Tarski 1956, 152–278..

Tarski, A. 1933, "Der Wahrheitsbegriff in den formalisierten Sprachen", *Studia Philosophica* I, 261–405.; repr. in Tarski 1986, Vol. 2, 51–198.

Tarski, A. 1936, „Grundlegung der wissenschaftlichen Semantik", in *Actes du Congrés International du Philosophie Scientifique*, vol. 3, Hermann, Paris, 1-8, Eng. tr. in Tarski 1956, 401–408.

Tarski, A. 1941, *Introduction to Logic and to the Methodology of Deductive Sciences*, Clarendon Press, Oxford.

Tarski, A. 1944, "The Semantics Definition of Truth and the Foundations of Semantics", *Philosophy and Phenomenological Research* 4, 341–375; repr. in Tarski 1986, Vol. 2, 661–699.

Tarski, A. 1956, *Logic, Semantics, Mathematics. Papers from 1923 to 1938*, Clarendon Press Oxford, 2nd ed., Hackett, Indianapolis 1983.

Tarski, A. (1969). Truth and proof. *L'age de la Science, 1*, 279–301; repr. Tarski 1986, Vol. 4. 399–422.

Tarski, A. 1986, *Collected Papers*, Vols. 1–4, Birkhäuser, Basel.

Tarski A. 1986a, "What Are Logical Notions?", *History and Philosophy of Logic* 7, 143–154.

Tarski, A., Mostowski, A. and Robinson, R. 1953, *Undecidable Theories*, North-Holland, Amsterdam.

Taylor, R. 1962: "Fatalism", *Philosophical Review* 71, 56–66.

Tegtmeier, E. (ed.) 2012, *Studies in the Philosophy of Herbert Hochberg*, Ontos, Frankfurt am Main.

Thomason, R. 1986, "Paradoxes and Semantic Representation", in Halpern 1986, 225–239.

Tkaczyk, M. 2007, "Is the Ontological Proof of God's Existence and Ontological Proof of God's Existence", *Logic and Logical Philosophy* 16, 289–309.

Tomasello M. 2010, *Origins of Human Communication*, The MIT Press: Cambridge, Mass.

Thomson, J. J. 2008, *Normativity*, Open Court, LaSalle.

Troelstra, A. S., van Dalen, D. 1988, *Constructivism in Mathematics. An Introduction.* North-Holland, Amsterdam.

Turner R. 1990, *Truth and Modality for Knowledge and Representation*, Pitman, London.

Turner, R 1990a, "Logics of Truth", *Notre Dame Journal of Formal Logic* 31, 308–329.

Twardowski, B., Woleński, J. (eds.) 1994, *60 Years of Tarski's Definition of Truth*, Philed, Kraków

Twardowski, K. 1902, 'Über sogennante relative Wahrheiten', *Archiv für systematische Philosophie* 8, 415–447; Eng. tr. in Twardowski 1999, 147–169.

Twardowski, K. 1999, *On Actions, Products and Other Topics in Philosophy*, Rodopi, Amsterdam 1999.

Urchs, M. 1992: "On Determinism – Still Unproved in Classical Logic", *Bulletin of the Section of Logic* 21, 168–170.

Van Heijenoort, J 1967, "Logic as Calculus and Logic as Language", *Synthese* 17, 324–330; repr. in Van Heijenoort 1985, 11–16.

Van Heijenoort, J. (ed.) 1967a, *From Frege to Gödel. A Source Book in Mathematical Logic, 1879–1931*, Harvard University Press, Cambridge, Mass.

Van Heijenoort, J. 1985, *Selected Essays* Bibliopolis, Naples.

Veldman, W. 1976, "An Intuitionistic Completeness Theorem for Intuitionistic Predicate logic, *Journal of Symbolic Logic* 41, 159–166.

Vuellimin, J. 1996, *Necessity or Contingency. The Master Argument*, CSLI, Stanford.

Waismann, F. 1956:,"How I See Philosophy?", in Lewis 1956, 345–380; repr. in Waismann 1968, 1–38.

Waismann, F. 1965, *The Principles of Linguistic Philosophy*, Macmillan, London.

Waismann, F. 1968, *How I See Philosophy*, Macmillan, London.

Waismann, F. 1976, *Logik, Sprache, Philosophie*, Stuttgart: Reclam.

Wedgwood, R. 2007, *The Nature of Normativity*, Clarendon Press, Oxford.

Weir, A. 2010, *Truth through Proof. A Formalist Philosophy of Mathematics*, Clarendon Press, Oxford.

Weingartner, P. 1979, "Ens et verum convertuntur", *Freiburger Zeitschhrift für Philosophie*, XXVI, 1979, 145–162.

Weingartner, P., 2000, *Basic Questions of Truth*, Kluwer, Dordrecht.

Wereński S. 2007, *Topologia* (Topology), Wydawnictwo Politechniki Radomskiej, Radom.

Wessel, H. 1999: *Logik und Philosophie*. Λογος, Berlin.

White, A. 1982, *The Nature of Knowledge*, Rowman & Littlefield, Totova.

Windelband, W. 1912, *Die Prinzipien der Logik*, J. C. B. Mohr (Paul Siebeck), Tübingen.

Wittgenstein, L. 1922, *Tractatus Logico-Philosophicus*, Kegan, Trench, Trubner, London.

Wittgenstein L. 1953, *Philosophical Investigations*, Blackwell, Oxford.

Woleński, J. 1994, "Is Tarski's Conception of Truth Relativistic?", in Twardowski, Woleński 1994, 96–12.

Woleński, J. 1995, "Logic and Falsity", in Childers, Majer 1995, 95–105.

Woleński, J. 1997, "Two Theories of Transcendentals", *Axiomathes* 8, 367–380.

Woleński, J 1997a, "Husserl and the Development of Formal Semantics", *Philosophia Scientiae* l.4,151–154; repr. in Woleński 1999, 110–114.

Woleński, J. 1998, "Truth and Bivalence", *The Logica Yearbook 1997*, Filosofia, Praha 1998, 36–43.

Woleński, J. 1999, *Essays in the History of Logic and Logical Philosophy*, Jagiellonian University Press, Kraków.

Woleński, J. 1999, "Semantic Conception of Truth as a Philosophical Theory", in Peregrin 1999, 51–65.

Woleński, J. 2001, "In Defence of the Semantic Definition of Truth", *Synthese* 126, 67–90.

Woleński, J. 2003, "Polish Attempts to Modernize Thomism by Logic (Bocheński and Salamucha)", *Studies in East European Thought* 55, 299–313.

Woleński, J 2003a, "Truth and Satisfaction by the Empty Sequence", in Rojszczak, Cachro, Kurczewski 2003, 267–276; repr. in Woleński 20011, 17–24.

Woleński, J. 2004, "Gödel, Tarski and Truth", *Revue Internationale de Philosophie* 59, 459–490; repr. in Woleński 2011, 101–123.

Woleński, J. 2004a, "First-Order Logic: (Philosophical) Pro and Contra", in Hendricks, Neuhaus, Pedersen, Scheffler, Wansing 2004, 369–399.

Woleński, J. 2004b. "What is Formal in Formal Semantics". *Dialectica* 58, 427–436; repr. in Woleński 2011, 81–89.

Woleński, J. 2004c, "On Some Formal Properties of Truth", in Kolak, Symons 2004, 91–100, repr. in Woleński 2011, 91–100.

Woleński, J. 2004d, "Analytic vs. Synthetic and A Priori vs. A Posteriori", in Niniiluoto, Sintonen, Woleński 2004, 781–839.

Woleński, J. 2005, „O zdaniach identycznościowych" (On Identity Statements), *Acta Universitatis Wratislaviensis* – 399. *Logika* 23, 175–195; rep. in Woleński 2017, 7–34.

Woleński, J. 2006, "A Generalization of Hume's Thesis", *Philosophia Scientiae* 10(2006), 111–115; repr. (with additions) in Woleński 2011, 155–162.

Woleński, J. 2006a, "Three Contributions to Logical Philosophy", in Jadacki, Paśniczek 2006, 195–213.

Woleński, J. 2008, "Read about T-scheme". Rahman, Tulenheimo, Genot 2008, 199–203.

Woleński, J. 2008a, "Remarks on Identity across Possible Worlds", *Bulletin of the Section of Logic* 37(3/4), 273–287; repr. in Woleński 2011, 243–253

Woleński, J. 2008b, "Applications of Squares of Oppositions and Their Generalizations In Philosophical Analysis", *Logica Universalis* 2(1), 13–29; repri. in Woleński 2011, 255–269.

Woleński, J. 2008c, "The Character of T-Equivalences", in Dégremont, Keiff Rückert 2008, 511–521; repr. in Woleński 2011, 231–241.

Woleński, J. 2009, "Theism, Fideism, Atheism, Agnosticism, in Johansson, Österberg, Śliwiński 2009, 512–521.

Woleński, J. 2009a, "The Principle of Bivalence and Suszko Thesis", *Bulletin of the Section of Logic* 37 (3/4), 99–110; repr. in Woleński 2011, 285–292.

Woleński, J. 2010, "Meaniungfulness, Meaninglesness and Language–Hiererchies: Some Lessons from Ingarden's Criticism of the Verifiability Principle", *Polish Journal of Philosophy* IV (2), 35–47; repr. in Woleński 2011, 313–323..

Woleński, J 2011, *Essays on Logic and Its Apllications in Philosophy*, Peter Lang, Frankfurt am Main.

Woleński, J. 2012, Logic as Calculus versus Logic as Language, Language as Calculus versus logic as Universal Medium and Syntax versus Semantics. *Logica Universalis*, *6*, 587–596; repr. in this volume.

Woleński, J. 2012, "Truth is Eternal if and only if it is Sempiternal", in Tegtmeier 223–230; repr. in this volume.

Woleński, Jan 2014 "An Abstract Approach to Bivalence", *Logic and Logical Philosophy* 23(1), 3–14; repr. in this volume.

Woleński, J. 2014, "Logic and Rule-Following", in Gizbert-Studnicki, Araszkiewicz, Bana, Płeszka 2014, 395–402.

Woleński, J. 2016, *Wykłady o naturalizmie* (Lectures on Naturalism), Wydawnictwo UMK, Toruń.

Woleński, J. 2017, *Logika i inne sprawy* (Logic and Other Matters), Copernicus Center Press, Kraków.

Woleński, I. (in preparation), *Semantics and Truth.*

Woleński, J. Simons, P.. 1989: "De Veritate: Austro–Polish Contributions to the Theory of Truth from Brentano to Tarski", in Szaniawski 1989, 391–453.

Wolter, A. B. 1946, *The Transcendentals and Their Fu ction in the Metaphysics of Duns Scotus*, The Catholic University of America Press, Washington.

Yablo, S. 1993, "Paradox without Self-Reference", *Analysis* 54(4), 251–252.

Yablo, S. 2006, "Circularity and Paradox", in Bollander, T., Hendricks, V. T., Pedersen, S. (eds.) 2006, 165–183.

Yockey H. 2005, *Information Theory, Evolution and the Origin of Life*, Cambridge University Press, Cambridge.

Zagzebski, L. 1991, *The Dilemma of Freedom and* Foreknowledge, Oxford University Press.

Index of Names

Malcev, A. I. 79, 90, 104, 111, 129-131, 139
Montague, R. 28

Y

Z

Index of Subjects

Polish Contemporary Philosophy and Philosophical Humanities
Edited by Jan Hartman

Vol. 1 Roman Murawski: Logos and Máthēma. Studies in the Philosophy of Mathematics and History of Logic. 2011.

Vol. 2 Cezary Józef Olbromski: The Notion of *lebendige Gegenwart* as Compliance with the Temporality of the "Now". The Late Husserl's Phenomenology of Time. 2011.

Vol. 3 Jan Woleński: Essays on Logic and its Applications in Philosophy. 2011.

Vol. 4 Władysław Stróżewski: Existence, Sense and Values. Essays in Metaphysics and Phenomenology. Edited by Sebastian Kołodziejczyk. 2013.

Vol. 5 Jan Hartman: Knowledge, Being and the Human. Some of the Major Issues in Philosophy. Translated by Ben Koschalka. 2013.

Vol. 6 Roman Ingarden: Controversy over the Existence of the World. Volume I. Translated and annotated by Arthur Szylewicz. 2013.

Vol. 7 Jan Hartman: Philosophical Heuristics. Translated by Ben Koschalka. 2015.

Vol. 8 Roman Ingarden: Controversy over the Existence of the World. Volume II. Translated and annotated by Arthur Szylewicz. 2016.

Vol. 9 Tomasz Kubalica: Unmöglichkeit der Erkenntnistheorie. Leonard Nelsons Kritik an der Erkenntnistheorie unter besonderer Berücksichtigung des Neukantianismus. 2017.

Vol. 10 Renata Ziemińska: The History of Skepticism. In Search of Consistency. 2017.

Vol. 11 Jan Woleński: Logic and Its Philosophy. 2018.

www.peterlang.com

www.ingramcontent.com/pod-product-compliance
Ingram Content Group UK Ltd.
Pitfield, Milton Keynes, MK11 3LW, UK
UKHW041825031026
15587UKWH00002B/560

9 783631 727454